T0239406

Novel Plant Imaging and Analysis

Tomoko M. Nakanishi

Novel Plant Imaging and Analysis

Water, Elements and Gas, Utilizing Radiation and Radioisotopes

 Springer

Tomoko M. Nakanishi
Graduate School of Agricultural and Life Sciences
The University of Tokyo
Bunkyo-ku, Tokyo, Japan

ISBN 978-981-33-4994-0   ISBN 978-981-33-4992-6   (eBook)
https://doi.org/10.1007/978-981-33-4992-6

This Springer imprint is published by the registered company Springer Nature Singapore Pte Ltd.
The registered company address is: 152 Beach Road, #21-01/04 Gateway East, Singapore 189721, Singapore

Preface

This book summarizes what kind of research I have been pursuing throughout my career as a researcher. Originating as a radiochemist, I jumped into the plant physiological field, which was filled with riddles. As I proceeded further in this field, I noticed that most of the questions I found are about the fundamental activity of the plant, for example, how ions are actually moving in the solution within a plant or the balance of intake and outgo. Although we could determine some mechanisms, such as the roles of transporters or channels, there was no explanation of the total movement of water or ions throughout a whole plant, rather than at specific sites of the cells. In addition, there are many questions, such as how the concentration of each ion changes with the movement of water. From a chemical point of view, diffusion and osmotic pressure, sometimes including Brownian motion, are the fundamental explanations of movement. However, it seemed that these were not the main theme to incorporate into the discussion. What regulates the movement of ions or water in each tissue, balancing with the other tissues? The research to determine these actual movements of water or ions could be slightly different from the research to determine the role or mechanism of transporters or channels. This research is to know the whole plant activity itself. Therefore, before observation at the microscopic level, the overall activity of the plant was investigated. In addition, some of our observations are now progressing to the microscopic level, including gene expression.

Another point I noticed was that by the most effective utilization of radiation or radioisotopes (RIs), the fundamental activity of the plant could be analyzed from different perspectives, which might lead us to develop an original research field. In recent decades, the number of researchers utilizing radiation or RIs has been drastically reduced. Most of researchers seem to avoid using RIs or radiation now. However, without utilizing these tools, we cannot study or determine many basic activities of plants. Since water and inorganic ions are essential for plant growth, I focused on these fundamental matters, developing nondestructive tools, mostly imaging methods.

Many questions about plant activities arose from our findings. For example, what is the chemical state of ions or water, which roots are absorbing from soil? Through neutron beam imaging, we found that there is always a space adjacent to the surface of the root, indicating that the root is not touching the solution in the soil. This is because of the root tip movement, circumnutation, which was confirmed by using a Super-HARP camera. The root was always pushing the soil aside to guide the orientation of the root development. The next question was whether the root was absorbing water solution or water vapor, and the same question could be asked for ions in soil. Further questions addressed metals. Is the root absorbing metal vapor in soil? I could not yet determine the answer to these questions. Another riddle is that, no one had ever discussed the circulation of water in the internode within a plant. When trace amount of water was measured by labeling water with RI, a tremendous amount of water was found to leak horizontally from xylem tissue and push the water already present in the internode into the xylem to travel upward via xylem tissue. In approximately 20 min, half of the water present in the internode was calculated to be replaced. Was there any previous measurement of these movements? Photosynthesis activity presented yet another basic riddle. It was amazing to determine that the transfer orientation of photosynthate was different according to the tissue where the photosynthate was produced. The motive force affecting this phloem flow orientation is not known. This phenomenon was found by imaging the $^{14}CO_2$ gas fixation process by a real-time RI imaging system we developed. The real-time imaging of RI movement provided another exciting method to analyze the routes of xylem flow and phloem flow.

To pursue original research, we developed our own measurement and imaging systems. Except for the neutron source, which is a research reactor, all the devices or systems presented in this book are our original work. My goal is to show how the utilization of radiation or RIs is an indispensable tool for plant research; therefore, this book mainly focuses on new imaging or measuring methods with the results obtained utilizing the tools presented, and most of the further research results we pursued are omitted. However, fundamental questions encountered in our research are described in many parts of this book.

Since we are sure that there are hardly any other comparable systems developed by other people now, including systems for the measurement of the actual movements of water and ions, the author sincerely hopes that this book will attract research interest in the utilization of radiation and RIs.

The following is a summary of each chapter in this book.

Bunkyo-ku, Tokyo, Japan   Tomoko M. Nakanishi

Acknowledgement

The studies presented in this book were mostly proceeded in my laboratory, named "Radio-Plant Physiology", Faculty of Agricultural and Life Sciences, The University of Tokyo, since I joined this faculty about 30 years ago from radiochemistry field and I found this name fits to my laboratory. To introduce the application of radiation and radioisotopes, I sincerely thank for the people who participated at each category of my work in the laboratory.

First of all, I thank to the people who are now in the Radio-Plant Physiology Laboratory, Prof. Keitaro Tanoi, Associate Prof. Natsuko I. Kobayashi, Assistant Prof. Ryosuke Sugita and the people who once worked with us and now in other organizations, Associate Prof. Satomi Kanno at Nagoya University, Associate Prof. Naoto Nihei at Fukushima University, Associate Prof. Atsushi Hirose at Hoshi University, Associate Prof. Jun Furukawa at Tsukuba University, Dr. Tomoyuki Ohya at QST (National Institutes for Quantum and Radiological Science and Technology), Associate Prof. Hiroki Rai at Akita Prefectural University, Dr. Masato Yamawaki at AIST (National Institute of Advanced Industrial Science and Technology). And I am much obliged to all the people who worked at the laboratory, as secretaries, Ms. Utako Shinohara and Ms. Megumi Anzai, technicians, students, graduate students.

I am also thankful to the people at JAEA (Japan Atomic Energy Association) who arranged the use of an atomic reactor or an accelerator with very useful discussions, Drs. Masato Matsubayashi, Hiroshi Iikura and other staffs and I would like to express my gratitude to Drs. Kazutoshi Suzuki, Kotaro Nagatsu, Noriko Nishioka and other staffs at QST who arranged for us to produce ^{18}F, ^{15}O as well as ^{28}Mg. I especially thank to Dr. Ren Iwata who is a Prof. Emeritus at Tohoku University now to provide us the chemical separation method of ^{28}Mg from the target. About phosphate imaging, I would like to thank deeply to Dr. Laurent Nussaume at CEA, in France, for the corporative work using RRIS (Real-time Radioisotope Imaging System).

Introduction and Executive Summary

Currently, powerful methods derived from molecular genetics have resulted in a tendency to focus research on the molecular aspects of biology and tend to leave behind important aspects of the activity of intact plants. However, the intact plant itself has high potential to integrate functions and to respond to diverse environmental conditions. To study the activity and development of living plants, nondestructive techniques are basic and essential. The imaging method is a particularly important tool. Fluorescent imaging is rapidly developing and has become overwhelmingly common in most biological studies. However, imaging utilizing radiation or radioisotopes (RI) has a definite advantage compared to fluorescent imaging from the perspective of quantitative analysis and indifference to lighting conditions. All of the data presented here introduce our work in water- and element-specific imaging and measurement in plants. However, a very limited number of people have utilized radiation or radioisotopes (RIs) for physiological research in living plants. Therefore, I would like to present my experience showing how imaging using radiation or RI holds promise to open new fields of plant research.

Although water and elements are essential for plant growth, kinetic studies of these two materials in situ are not progressing well.

In the case of water, for example, when we examine a representative activity of the plant, photosynthesis, where it is well known that sugar is produced from water and carbon dioxide gas in the air, we soon notice that there is very little research on water. There have been many studies on the mechanism of photosynthesis, the chemicals produced after carbon fixation, or the effect of carbon dioxide gas. However, in most cases, it is taken for granted that there is sufficient water already present in plant tissue for the chemical reaction. Though we could determine the kinds of chemicals produced by photosynthesis, we do not know how much water is needed or moving within the plant.

Water

Our first target was water, namely, how to obtain a water-specific image nondestructively. Neutron beam imaging was applied to provide water-specific images. Using a neutron beam, we could visualize water-specific images of plants, including roots and flowers, which were never shown before. Each image suggested the plant-specific activity related to water.

However, only a small number of people employ water-specific imaging produced by neutron beams for water-related studies in plants. Therefore, we briefly present how to acquire the image and what kind of water image is taken by neutron beam irradiation. We present a variety of plant samples, such as flowers, seeds, and wood disks. It was noted that neutrons could visualize the roots imbedded in soil without uprooting. When a spatial image of the root imbedded in soil was created from many projection images, the water profile around the root was analyzed. Then, fundamental questions were raised, such as whether plants are absorbing water solution or water vapor from the soil, because there was always a space adjacent to the root surface and hardly any water solution was visualized there. The roots are in constant motion during growth, known as circumnutation, and it is natural that the root tip is always pushing the soil aside to produce space for the root to grow. If the roots are absorbing water vapor, then the next question is about metals. Are the roots absorbing metal vapor? Since we tended to employ water culture to study the physiological activity of plants, the physiological study of the plants growing in soil was somewhat neglected. Later, when we could develop a system to visualize the movement of element absorption in a plant, there was a clear difference in element absorption between water culture and soil culture.

The next approach to research on water was to measure the small amount of water actually moving within a plant. The best method is to utilize radioisotope (RI)-labeled water and measure the radiation from outside of the plant. However, it is rather difficult to label water, since there are only limited kinds of RI for tracing water. The first trial utilized ^{18}F because trace amounts of this nuclide are produced when water is irradiated with a helium beam. It is well known that trace amounts of RI exhibit radiocolloidal behavior and move with larger amounts of chemicals. When utilizing ^{18}F to trace water movement, another fundamental question to consider was the features that characterize drought-tolerant and drought-sensitive plants. It is natural to suppose that drought-tolerant plants have strong water absorption; therefore, by analyzing the water absorption mechanism of tolerant plants and by introducing this function to sensitive plants, it might be possible to make the sensitive plants more tolerant.

However, when water uptake was studied in naturally developed drought-tolerant and drought-sensitive cowpea, selected from 2000 cowpea plants grown in the field of Africa, the result was unexpected. Under normal conditions, the amount of water absorbed by the drought-tolerant strain was much lower than that absorbed by the sensitive strain, as if showing the low capability of water absorption. On the other hand, much higher amount of water was absorbed by the sensitive sample. However, this water absorption amount drastically changed after drying treatment. The tolerant strain began to absorb much more water than usual, whereas the sensitive strain could not absorb as much water as before the treatment. That is, the drought-tolerant plant required only a small amount of water under normal conditions, but when a drought condition was introduced, some mechanism was activated to absorb much more water. This result provided us with an important lesson. Analyzing the mechanism of drought tolerance only by comparing the water absorption of tolerant and sensitive plants might not readily reveal the reason for drought tolerance. The features of the naturally produced plants showed us different mechanisms that might not match our expectations developed in the laboratory.

Next, we performed water measurements using ^{15}O-labeled water, which has an extremely short half-life of 2 min. Here, we found another astonishing result, which was "water circulation" in the plant internode. A tremendous amount of water was always leaking from xylem cells, which had been regarded as a mere pipe to transfer water from the root to the aboveground parts. Although nondestructive measurements could not be performed, ^{3}H-labeled water was also employed to verify the phenomenon of horizontal water leakage from xylem cells. In another subsequent study, it was shown that the water flowing out from the xylem was pushing out the water already present in the stem and then returning to the xylem again to move upward. The water velocity in the internode was kept constant, and through simulation, it took less than 20 min to exchange the water already present in the stem with newly absorbed water. How does this happen? Is the water already present in the internode different from newly absorbed water? Since the half-life of ^{15}O is extremely short, the measuring system we developed needed special devices to determine the small amount of water actually moving in the stem.

Elements

Since the elements absorbed from roots are moving with water, studying element-specific movement within a plant is another theme of this book. For the first stage of the study of the elements, the distribution of the element within the plant tissue was presented employing neutron activation analysis (NAA). Since NAA allows nondestructive analysis of the elements in the sample, this is the only method to measure the absolute amount of elements in the sample. With the extremely high sensitivity to heavy elements and allowing multielement analysis, NAA provided the content of the many elements in each plant tissue.

The results showed that the element-specific profile varied throughout the whole plant, and this distribution tendency remained similar throughout developmental stage. There were many junctions of element-specific concentrations between the tissues, suggesting barriers to the movement of the elements. Generally, heavy elements tended to accumulate in roots, except for Mn and Cr. Even in a single leaf, there existed an element-specific concentration gradient. Of the elements measured,

Ca and Mg showed changes in concentration with the circadian rhythm. Since the amount of the element in a plant reflects the features of the soil where the plant grows, multielement analysis of the plant could specify the site of the agricultural products produced.

Before addressing the development of a real-time RI imaging system (RRIS), the production of RIs for essential elements for plant nutrition, ^{28}Mg and ^{42}K, is presented. The reason why concentrating on RIs is because when we examine the history of plant research, physiological research on the elements without available radioisotopes has not been well developed. For example, the boron (B) transporter was recently found, whereas there have been many transporter studies on phosphate (P), which has available radionuclides, ^{32}P and ^{33}P. Since there are no available radioisotopes for B for use in experiments, the study of B in plants is far behind compared to the other elements. Although it was not possible to produce radioactive nuclides for B, efforts were made to image other important elements for plant nutrition. Therefore, we developed a preparation method for elements whose available RIs were not previously employed in plant research. ^{28}Mg and ^{42}K are the radioisotopes we prepared; ^{28}Mg was produced using an accelerator, followed by chemical separation of the nuclide from the target. A root absorption study using ^{28}Mg as a tracer is presented as an example. It was found that the orientation of Mg transfer was different according to the site of the root where Mg was absorbed.

Mg and Ca are in the same group in the periodical table, have similar chemical properties, and are stained simultaneously by fluorescence probes; therefore, it was difficult to distinguish one from the other. Ca research is much further advanced than Mg research because of the high contribution of the available radioactive nuclide ^{45}Ca. Developing research on Ca has induced the development of fluorescent staining methods. However, since the overwhelming amount of Ca makes it difficult to distinguish Mg and Ca by imaging, the specific role of Mg has not yet been clarified.

Real-Time RI Imaging of Elements

Then, we proceeded to develop an imaging method utilizing the available RIs. We developed two types of real-time RI imaging systems (RRIS), one for macroscopic imaging and the other for microscopic imaging. The principle of visualization was the same, converting the radiation to light by a Cs(Tl)I scintillator deposited on a fiber optic plate (FOS). Many nuclides were employed, including ^{14}C, ^{18}F, ^{22}Na, ^{28}Mg, ^{32}P, ^{33}P, ^{35}S, ^{42}K, ^{45}Ca, ^{48}V, ^{54}Mn, ^{55}Fe, ^{59}Fe, ^{65}Zn, ^{86}Rb, ^{109}Cd, and ^{137}Cs. In the case of macro-RRIS, the steps to develop the imaging system were as follows. In the first generation, the plant sample and all the imaging devices had to be kept in the dark. In the second generation, a sealed plant box was prepared where only the aboveground part of the plant was able to undergo light irradiation to protect the highly sensitive CCD camera. Then, an on/off switch to irradiate the sample was introduced into the imaging box, and a plastic scintillator was tested for imaging a large-scale plant.

Since radiation can penetrate the soil as well as water, the difference between soil culture and water culture was visualized. The plants grew much faster in water culture but with a low yield of grain. These culturing methods were applied to show the difference in ^{137}Cs absorption by the rice plant. ^{137}Cs was hardly absorbed by rice roots growing in soil, whereas water culture showed high absorption, which could provide some reassurance after the Fukushima Nuclear Accident and could indicate an important role of soil in firmly adsorbing the radioactive cesium.

^{28}Mg and ^{42}K, whose production methods were presented, were applied for RRIS to visualize the absorption image from the roots. In addition to ^{28}Mg and ^{42}K, many nuclides were applied to image absorption in the roots. Each element showed a specific absorption speed and accumulation pattern. There were three types of movement velocity: fast, medium, and slow. The image analysis of the absorption of Mg is presented as an example. Through successive images of the element absorption, phloem flow in the aboveground part of the plant was analyzed. The element absorption was visualized not only in the roots but also in the leaves, a basic study of foliar fertilization.

In the case of the microscopic imaging system, a fluorescence microscope was modified to acquire three images at the same time: a light image, fluorescent image, and radiation image. Although the resolution of the image was estimated to be approximately 50 μm, superposition showed the expression site of the transporter gene and the actual ^{32}P-phosphate absorption site to be the same in Arabidopsis roots.

Imaging of ^{14}CO$_2$ Gas Fixation

We targeted not only the elements we can supply to the nutrient solution but also carbon dioxide gas to visualize the fixation process and the movement of assimilated carbon in a plant. This is another highlight of our study using RRIS. The interesting result was that the route of assimilated carbon was different depending on where the fixation took place. In Arabidopsis, most of the metabolites after photosynthesis were transferred to the tip of the main internode and roots when ^{14}CO$_2$ gas was fixed and photosynthates were produced at rosette leaves, whereas most of the metabolites moved to the tip of the branch internode and hardly moved down to the roots when ^{14}CO$_2$ gas was supplied to the aboveground parts of the plant other than rosette leaves. However, the route was also dependent on the developmental stage of the plant. Interestingly, it was possible to visualize and trace which tissue performed the fixation of ^{14}CO$_2$ gas, i.e. carbon could be traced from the fixation site in tissue to tissue formation. However, especially in the case of ^{14}C imaging, image analysis should be carefully performed because of the self-absorption of the β-rays in tissue. To image ^{14}CO$_2$ gas fixation in larger samples, approximately 50 cm in height, a plastic scintillator was introduced, and the assimilation process of the gas was visualized for rice and maize.

3D Image Construction and MAR

Finally, spatial (3D) image construction from imaging plate (IP) images and the development of the microautography (MAR) method were presented. A rice grain was sliced every 5 μm, IP images were taken for successive slices, and the series of 2D images acquired by an IP were used to construct 3D images. In the case of ^{109}Cd and ^{137}Cs, the spatial distributions in the grain showed that the concentrations increased at the surface of the grain during the maturing process.

MAR was developed to increase the resolution of the image for the sliced plant sample. This revised MAR method showed the detailed distribution of ^{137}Cs accumulation in embryos, which indicated that the plumule and radicle, which grow as a meristem of a root or shoot, were protected from the accumulation of heavy elements. Because of daguerreotype imaging, MAR is now hardly used, and the film emulsion is not available. Here, this method was essentially recreated with a revised processing method for both sample preparation and imaging process.

Contents

About the Author

Tomoko M. Nakanishi The author is Prof. Emeritus and Project Prof., The University of Tokyo as well as President of Hoshi University in Tokyo. She majored in radiochemistry and received Ph.D. from The University of Tokyo, where she had set up her laboratory named radio-plant physiology. She was interested in role of water and elements in plants and found new aspects of plant physiology never expected before utilizing radiation and radioisotopes. After Fukushima nuclear accident, her group has vigorously studied the agricultural consequences of radioactive contamination. She received George Hevesy Medal Award, Ordre national du Mérite from French president and elected as a Foreign Member of the Royal Swedish Academy of Engineering Sciences and the Royal Society of Arts and Sciences in Gothenburg. She received an honorary Doctor of Chalmers University of Technology in Sweden.

Part 1
Water in a Plant

Water is essential for plants to grow, but we have been taking it as natural that water is always present in plants without considering deeply how much water is actually needed or moving during living plant activities. Although water movement plays an important role both in chemical processes and in forming the physical structure of a plant, it has not been studied in detail, mainly because the tools for such research are lacking.

There are several ways to trace water movement in plants. Dyes such as fuchsin were used to visualize water pathways in living plants, illustrating the water movement by the color movement. However, the movement of a dye dissolved in water must be different from that of water itself, and it was difficult to analyze the amount of water actually moving by relying on the color. It was considered that nondestructive techniques for analyzing water within the living cells or tissues of plants could be extremely important tools for studying plant activity.

To acquire water images in a plant nondestructively, computer-assisted tomography with X-ray attenuation measurement has been reported, although the resolution of the method was not high enough to acquire water-specific images. Another promising approach to obtain water images is NMR. However, the resolution of the images is theoretically restricted to not less than 10 μm because of relaxation time, and the size of the sample is limited. In most cases, for NMR, the plant sample must be in a test tube where the environmental conditions, such as temperature, light, or humidity, are difficult to adjust. The last candidate nondestructive method to image water is applying a neutron beam. However, to utilize the neutron beam, special equipment, such as an atomic reactor or an accelerator, is needed. To acquire a wide area of parallel neutron beams with high intensity, collimated neutrons from an atomic reactor are preferable to those obtained using an accelerator, where the intensity of the beam for imaging is lower than that of the reactor and the focused beam has to move to scan the target. In the case of neutron beam imaging, there is no theoretical limit of the resolution in the image. The resolution is dependent on the method of obtaining the image. For example, when an X-ray film is applied, the size of the Ag grain coated on the surface of the film determines the resolution, approximately 20 μm.

Although neutron beam imaging provides static water images, when successive images are taken over time, it is possible to analyze water movement during plant growth. Since plant development is rather slow, in units of minutes or hours, when neutron images of the sample are taken periodically, the difference among the images could provide water movement. However, to acquire the on-time movement of water, not depending on statistical images, other methods are needed. For this purpose, to visualize the actual movement of water within a plant, we employed radioisotope-labeled water as a tracer. The actual water movement is presented in the next chapter (Chap. 2).

Chapter 1
Water-Specific Imaging

Keywords Nondestructive water image · Neutron beam image · Water-specific image · Root image in soil · 3D water image · 3D root image · Water in flower · Water in wood disk · Water absorption in seeds

1.1 Neutron Beam Imaging

When collimated neutrons are used to irradiate the sample, the neutrons penetrating the sample can produce images on film or transmit them to a computer through a CCD camera based on the intensity of the neutrons. The neutron beam image can be expressed as a kind of shade if it can be compared to shade images of light. To obtain a high-resolution neutron image, a parallel beam with high quality is needed, which is primarily decided by the features of the neutron collimator. When the length of the collimator is L and the aperture size of the collimator is D, the resolution of the image is reciprocal to L/D. To increase the resolution, it is preferable to use collimators with an L/D of more than 100. Another way to increase the resolution is to reduce the distance between the sample and the film or a camera. The resolution of the imaging time is dependent on the neutron flux and reciprocal to $(L/D)^2$. Another factor that can deteriorate the resolution is movement of the sample itself during the imaging. In the case of plant samples, the development of the tissue is not so fast; therefore, the imaging time is not as critical compared to the case of samples of engineering substances, such as imaging the formation of bubbles in the hot water in a metal pipe.

The features of a neutron beam compared to that of an X-ray are shown in Fig. 1.1, where the neutron attenuation coefficient, indicating the difficulty in penetrating the elements, is plotted in a log-scale manner. Generally, the penetration rate of metals is high for neutrons, and a small amount of neutrons could penetrate the light elements, which overall is roughly the reverse of that of an X-ray.

Supplementary Information The online version of this chapter (https://doi.org/10.1007/978-981-33-4992-6_1) contains supplementary material, which is available to authorized users.

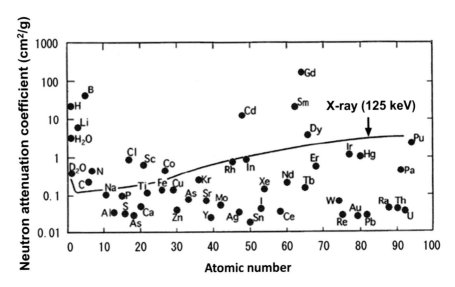

Fig. 1.1 Thermal neutron attenuation coefficient

As shown in Fig. 1.1, the attenuation coefficient of some light elements, including hydrogen, as well as rare earth elements, is extremely high, 100–1000 times higher than those of the other common elements. Because of the selective nuclear reaction of the neutron beam, when the neutron beam is used to irradiate the samples, including light elements or rare earth elements, the beam intensity is drastically reduced after penetrating the sample, which resulted in a clear image of these elements in the sample. In the case of X-rays, the beam reacts with electrons occupying the outside of the nuclide of the element. Since the number and density of the electrons increase with the atomic number of the element, the reaction of the X-ray with electrons increases, which results in a higher attenuation coefficient. Therefore, in the case of X-rays, this attenuation coefficient changes by factors, not by order of magnitudes, from lighter elements to heavier elements, which makes it difficult to distinguish the image of a specific element from that of the neighboring other elements.

Then, what can we see inside a plant irradiated with a neutron beam? In addition to hydrogen, there are only minimal amounts of lithium, boron or rare earths to attenuate neutron beams in plants. Therefore, the neutron beam can be regarded as producing a hydrogen image selectively. Since more than 80% of a living cell consists of water, the hydrogen image from plants is almost entirely water. After complete dehydration, there was hardly any neutron image in most of the plant samples. Therefore, we could regard the neutron image as a water-specific image. The acquired neutron image not only showed water distribution but also indicated the morphological development of the tissue itself. The following section shows how neutron beam images of intact plant tissues can give exquisite microscopic images never seen before.

1.2 Water-Specific Images by Neutron Beam

As written above, when a plant is irradiated with thermal neutrons, the neutron attenuation coefficient changes with the water content in the plant, producing different exposure images on an X-ray film. The different exposure images exhibit different levels of whiteness in an X-ray film; therefore, the whiteness of the film indicates changes in the water content in the tissue. This means that the water content can be estimated from the extent of whiteness in the image.

Figure 1.2 shows a schematic illustration of neutron radiography. When the sample was irradiated with a neutron beam, the neutrons penetrated after the sample produced an image on the X-ray film. The first trial used this kind of facility so that in the case of the plant sample, the sample had to be grown in a thin container to minimize the superposing water image around the roots.

The preparation of the sample and the irradiation method are described below with a soybean plant as an example. To obtain the neutron image of roots imbedded in soil, a soybean seedling was transplanted to a thin aluminum container (50 × 150 mm, 3 mm in thickness) packed with Toyoura's standard sand (silt type, 197–203 μm, in pore size) containing 15% (w/w) water. The upper part of the container was covered with aluminum tape to prevent water loss due to evaporation. The samples were kept at 26 °C under 70% humidity with 20,000 lux of light

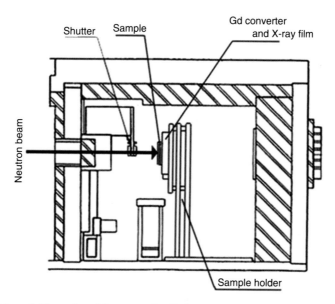

Fig. 1.2 Schematic illustration of the neutron irradiation chamber for neutron radiography [1]. The neutron beam from a reactor was irradiated from the left side to the sample, which was fixed on the cassette. After penetrating the sample, the neutrons were converted to radiation by a Gd n/γ converter set before an X-ray film in a cassette. Then, by developing the X-ray film, the image produced on the X-ray film was acquired

Fig. 1.3 Gadolinium converter and a cassette. The gadolinium (Gd) converter was prepared by depositing gadolinium on an aluminum plate (25 μm in thickness) and coated with sapphire to prevent oxidation. When irradiated with neutrons, gadolinium emits β-rays (conversion electrons) and γ-rays. The X-ray film is mainly exposed to β-rays to produce the image. The Cd converter and an X-ray film were placed in a cassette and sealed in vacuum

in a growth chamber. The plant sample was fixed to an aluminum cassette where a gadolinium n/γ converter and an X-ray film were sealed in vacuum (Figs. 1.3 and 1.4). The gadolinium converter was prepared by depositing gadolinium on an aluminum plate (25 μm in thickness) and was coated with sapphire to prevent oxidation. When irradiated with neutrons, gadolinium emits β-rays (conversion electrons) and gamma-rays, and the X-ray film is exposed mainly by the β-rays to produce the image. Therefore, the emulsion side of the film was placed tightly on the gadolinium side of the converter in the cassette.

The cassette was set vertically and was irradiated by thermal neutrons from an atomic reactor, JRR-3 M, installed at the Japan Atomic Energy Agency (JAEA) (Fig. 1.2). The total neutron flux was 1.9×10^8 n/cm^2.

After irradiation, the aluminum container became radioactive due to ^{28}Al formation, with a half-life of 2.3 m. However, after 15 m, the radioactivity of the sample was reduced to the background level. The plant samples themselves did not become radioactive after irradiation because the total irradiating neutron flux was relatively low. After irradiation, the film was developed, and the image was scanned by a scanner. When these irradiated plants were left to continue their growth, there was no observable effect compared with unirradiated plants.

1.2.1 2-Dimensional Images of Roots

When the growth of roots imbedded in soil could be visualized, we expected to determine new plant activities, since plant physiology has been developed mainly

Fig. 1.4 Plant samples prepared for neutron radiography. Since the Al board does not produce figures by neutron irradiation, the samples were prepared on an Al board, and this board was fixed on the cassette by Al tape. In the case of a soybean plant, the samples were grown in a thin aluminum container (50 × 150 × 3 mm) where Toyoura's standard sand (silt type, 197–203 μm, in pore size) containing 15% (w/w) of water was packed. The sample was kept in a phytotron and periodically removed to perform neutron imaging

depending on water culture, where the nutritional conditions can be strictly defined. However, plants growing in the field are influenced by soil conditions, matrix features, nutrient conditions, water amount, etc. All of these conditions differ from place to place and depend on season or climate. However, the physiology of plants growing in the field is not well known, and some of the information is from an agricultural perspective, i.e., how to obtain high yield. Is there any relationship between the growth of the aboveground part and that of the root? Is there any preference regarding the physical condition of the soil for growth? There are many questions to be solved for the plants growing in soil from the perspective of plant physiology. The visualization of roots imbedded in soil could be visualized, provide clues to solve these questions. Therefore, we tried to visualize what we usually cannot see by utilizing neutrons. The following is an example of the root images in soil taken by a neutron beam.

Figure 1.5 shows a neutron image of a soybean plant in an aluminum container 3 mm in thickness. The whiteness in the figure corresponds to the water amount; therefore, the lighter part corresponds to regions where the water content is higher. Since the water content in the roots is much higher than that of the surrounding sand,

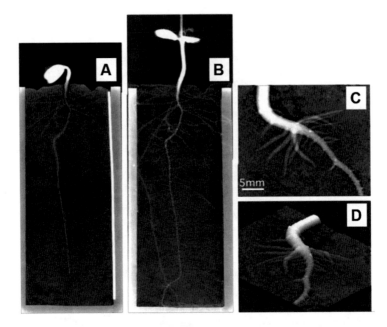

Fig. 1.5 Neutron images of soybean roots embedded in soil [2]. (**a** and **b**) the soybean plant images after 8 and 15 days. The whiteness in the figure corresponds to the water amount; therefore, the whiter part corresponds to the site with higher water content. Since the water concentration in the roots was higher than that in the soil, the morphological development of the roots and water movement were analyzed from the images. (**c**) magnification of the root when side roots emerged to develop. (**d**) conversion of the upper image to a 3-dimensional image, where the water amount was employed as the height

the root image is clearly shown in the figure even in the presence of soil. A water-deficient region, which was indicated as a darker site in the image, is clearly shown near the upper part of the main root.

The neutron image of the root imbedded in soil provides two pieces of information: the morphological development of the root and the water profile in the soil. As shown in the figure, we could visualize how the roots developed after the 8th day and 15th day of germination. The white dots shown in the main root indicate that a side root grew from this site and was superposed on the main root.

To discern the water content of the root more clearly, a part of the magnified root of the soybean plant 8 days after germination is shown. The three-dimensional image was acquired by processing the 2D figure of the root, and the height of the image corresponds to the whiteness and the amount of water. The 3D image showed more clearly that the water in the vicinity of the root, where the secondary root was developing, was taken up actively by the root, as shown by the clear emergence of the secondary root.

Figure 1.6 shows the morphological variety of soybean root images. Usually, balanced development of the lateral roots between both sides of the main root is

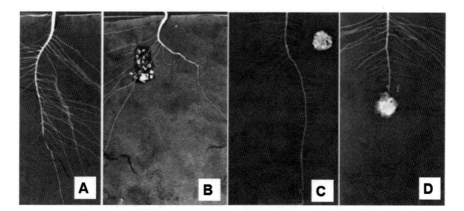

Fig. 1.6 Morphological differences of soybean roots embedded in soil. The balanced development of the lateral roots between both sides of the main root is commonly shown. However, when there is any change in soil condition on one side of the main root, the development of this side of the lateral roots changes from that of the other side. When there is some change near the root tip, the main root development ceases, but the lateral roots develop well to compensate for the ceased development of the main root. (**a**) control sample; (**b**) water-absorbing polymer containing water was placed on one side of the main root; (**c, d**) water-absorbing polymer containing 50 mM vanadium solution was placed on one side of the main root and in front of the root tip

shown. However, when there is any change in the soil condition in the vicinity of the main root, the development of one side of the lateral roots differs from that of the other side. For example, with enough water near the main root supplied from the water-absorbing polymer, the lateral roots on this side developed so well that some of them were even penetrating the water-absorbing polymer. On the other hand, when a heavy metal solution (water-absorbing polymer containing 50 mM vanadium) was placed on one side of the main root, the development of the lateral roots on this side ceased (Fig. 1.6c). In other cases, when this kind of polymer containing vanadium was placed just in front of the main root tip, the main root development stopped, but the lateral roots developed well, as if to compensate for the halted development of the main root (Fig. 1.6d).

These nondestructive images of root development in soil were applied to evaluate the effect of soil conditioning agents or water absorption polymers from the morphological development pattern of the roots or the change in water amount in soil. An earlier example is the effect of new lignin derivatives as soil conditioning agents in an acidic soil where aluminum ions inhibit plant growth1). Dissolved Al in soil is one of the major factors limiting crop production in acidic soil, which represents approximately 30–40% of the soil usable for agriculture in the world. To evaluate the effect of the soil conditioning agent under acidic soil conditions, the radish (*Raphanus sativus* var. *radicula* Pers.) was grown in soil containing lignin modified by radical sulfonation and alkaline oxygen treatment. Then, the root image obtained by the neutron beam was acquired. Figure 1.7 shows the traced images of the root acquired by neutron irradiation, and from these images, the root length was obtained

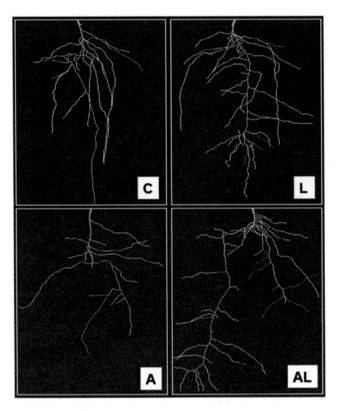

Fig. 1.7 Traced image of the radish roots acquired in neutron images [3]. The main root was colored blue, and the second and third roots were colored yellow and red. The effect of soil conditioning agents was observed from the morphological development pattern of the roots or the change in water amount in the soil. *C* control sample, *L* lignin derivatives (prepared as soil conditioning agents) were supplied to the soil A: Al ion was supplied to the soil (20 mM of Al ion solution was supplied to 16% of the soil), *AL* Al ion and lignin derivatives were supplied to the soil

and compared among different cases. The modified lignin was found to remove the toxic effect of aluminum ions and even improve the growth of the radish roots.

The second example of neutron imaging applied to the roots is the evaluation of water-absorbing polymers. In water-deficient areas, such as semiarid areas, how to keep water in the soil is a crucial problem, and sometimes special pottery containers filled with water are placed near the roots in the soil so that water gradually seeps out from the container and provides water to the plant for a considerable time. Therefore, water-absorbing polymers have attracted attention as replacements for such pottery containers. Since there are several kinds of water-absorbing polymers with different abilities to absorb water, two kinds of polymers were selected, and the method of supplying water from the polymer to the root was evaluated using neutron images.

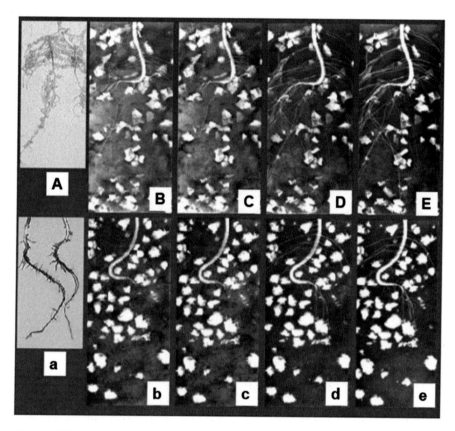

Fig. 1.8 Effect of two types of water-absorbing polymers on the soybean root development [4]. Two types of polymers were swelled in water solution and mixed with the soil: polyvinyl-alcohol polymer (**A**, **B**, **C**, **D**) and polyacrylic polymer (*a, b, c, d*). Then, soybean plants were grown in the soil mixture. In the case of the polyacrylic polymer, the water absorbed by the polymer was not supplied to the plant, and the color of the soil darkened with time, which indicates that the plant only absorbed water from the soil. The polyvinyl-alcohol polymer supplied water to the plants, and the image gradually faded. **A** and **B** are pictures of the roots grown in the container for a week

The neutron images of soybean root development in soil containing different kinds of water-absorbing polymer are shown in Fig. 1.8. The soybean seedlings were grown for 5, 6, 10, and 12 days in an aluminum container (150 × 70 × 3 mm) packed with soil containing polyacrylic water-absorbing polymer (*Acryhope*, Nippon Shokubai Kagaku Kogyo Co.) or polyvinyl alcohol copolymer (Mizumochi-ichiban, Nippon Gosei Kagaku Kogyo Co.). The white spots in the picture are the images of the polymers swelled with water. Therefore, the whiteness and size of each polymer showed where and how water was supplied from the polymer to the roots.

In the case of the polyacrylic water-absorbing polymer, as shown on the lower side of the images in Fig. 1.8, the size and whiteness of all the polymers were the same throughout the days of imaging, indicating that the water content in the polymer was not changed. That is, the plant absorbed water only from the soil, and the water in the polymer was not supplied to the plant. Then, a shortage of water occurred, and the roots did not develop well. The length of the main root was approximately the same in all the images.

The upper images in Fig. 1.8 are neutron images of a soybean seedling grown in soil containing polyvinyl alcohol copolymer. At first, there is much water in the soil, in contrast to the observations of the polyacrylic polymer, where water in the soil was absorbed quickly from the first stage. Then, the size of the polyvinyl alcohol copolymers, especially around the upper part of the root, decreased with root development. When the color of the soil and the polymer was compared, first, the soil became darker, and then the polymers became darker and decreased in size. This phenomenon indicated that water was supplied first from the soil and then from the polymer. The side root was well developed compared to that in the polyacrylic polymer in soil. After neutron images were taken, the roots were removed from the container, and the state of the roots was compared. As shown in Fig. 1.8, the roots, including side roots, did not grow well, and the color turned dark when polyacrylic polymer was applied, whereas when polyvinyl alcohol copolymer was supplied to the soil, the roots firmly penetrated the polymer, indicating that the roots searched for water and absorbed water from the polymer.

1.2.2   3-Diimentional Images of Roots

To study the water absorption of the root in more detail, a 3-dimensional (3D) neutron beam image was constructed. The neutron beam image was a 2D image as long as the neutron beam came from one direction of the collimator. When many 2D projection images are taken, with neutron irradiation applied to the sample from different angles, 3D images can be constructed from many 2D images through computer processing. Therefore, the plant was fixed on a rotating table, and the angle of irradiation to the sample was varied. Many 2D images were taken, and then a spatial image was constructed through a computer.

1.2.2.1   3D Image Construction

A five-day-old soybean seedling was transplanted to an aluminum container (35 mm φ, 150 mm) packed with Toyoura's standard sand containing 15% (w/w) water. The soil surface was covered with aluminum foil to prevent water loss due to evaporation. The samples were kept at 26 °C under 70% humidity with 20,000 lux of light in a growth chamber and were taken out periodically for imaging. These procedures were the same as those to acquire 2D images using an X-ray film. The exposure was performed in a JRR-3 M research reactor installed at the Japan Atomic Energy Research Agency (JAEA). The neutron flux used for the exposure was 1.5×10^8 n/ cm^2 s.

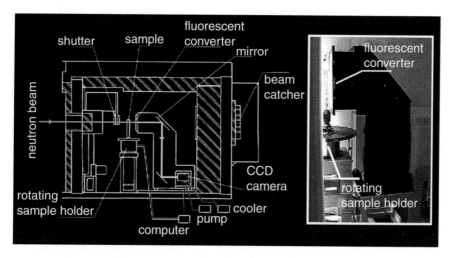

Fig. 1.9 Neutron irradiation chamber to acquire 3D image. Schematic illustration of the neutron irradiation chamber and a picture at JAEA. Neutrons were irradiated from the left-hand side, and the neutrons that penetrated the sample were converted to light by a fluorescence converter. The light beam was guided to a cooled CCD camera using two mirrors. The camera was shielded well with polyethylene and lead blocks. The sample was rotated by 180° in steps of 1°. The regulation of rotating the sample and the shutter speed of the camera were controlled using a computer

To obtain a CT image, the plant sample was set on a rotating disk (Fig. 1.9). The disk was fixed on a rotating table where a fluorescent neutron converter (NE426 equivalent made by Kasei Optonics Ltd.) was set as close as possible to the sample, at 2.2 cm from the rotation axis. The total neutron dose was 6.0×10^8 n/cm^2 per projection, and the L/D-value was 153. The shutter of a cooled CCD camera (C4880, Hamamatsu photonics, Co.) was opened for 4 s. The neutrons penetrating the sample were converted to photons by a fluorescent converter. The photon image was guided to the cooled CCD camera using two quartz mirrors. A MicroNikkor lens (F105 mm, Nikon Co.) was attached to the cooled CCD camera. The effective area of the fluorescent converter was approximately 5×5 cm. Since the CCD chip is very sensitive to scattered gamma-rays and neutrons, the camera was well shielded with polyethylene and lead blocks, each of which was $10 \times 10 \times 10$ cm. The sample was rotated in intervals of 1 up to 180°, and at each angle, a projection image was taken, i.e., 180 images were acquired for each sample to construct the CT image. The rotation of the sample and the shutter speed of the CCD camera were controlled by a computer. It took approximately 1 h to obtain 180 images, and each output image could be monitored on a computer. After neutron irradiation, approximately 1 h was needed to decay out the radioisotopes, mainly ^{28}Al (half-life: 2 min), produced during irradiation.

The size of the image transferred to the computer was 1000×1018 pixels. Then, each figure was corrected using two background images. One of the background images was taken when neutrons were absent (dark current), and the other was the image without the sample (shading). Therefore, at each image, dark current image

Fig. 1.10 Spatial image construction of a soybean root embedded in soil. A soybean plant grown in an aluminum pipe (30 mm φ × 150 mm) packed with soil was rotated by 180° in steps of 1°; at each angle, neutron imaging was performed. From 180 projection images, a spatial image was constructed using a computer

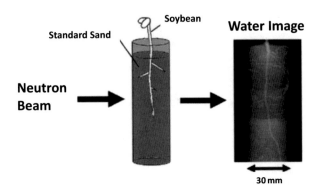

subtraction and shading correction were performed. Then, the root image, 600 × 1018 pixels, was cut out from the corrected figure to reconstruct the CT images. After sectional CT images were constructed at several root heights, sagittal CT images that included the main root were also constructed from a set of sectional CT images.

1.2.2.2   Water Movement Around the Root

Since plant development is slow, CT images can be used to evaluate the "static" spatial water distribution inside and around roots. The following is the spatial water distribution constructed from CT images in a soybean plant.

Successive neutron images of the soybean plant roots were taken during growth from 3 to 6 days. Each time, 180 projection images, acquired from different angles of irradiation to the sample, were processed on a computer to construct a CT image of the sample (Figs. 1.10 and 1.11). When successive sectional CT images at every pixel height were reconstructed and shown from the upper part to the bottom part of the container, they presented us with a basic question. Figure 1.12 shows the successive CT images. The main root is shown in the middle of all the images as a white spot, and whenever the side roots grow, they grow toward the wall of the container and then stop; therefore, the white dots at the wall are the tip of the side roots. The difference in the whiteness of the image showed the change in the water amount in the soil. However, it was always shown that surrounding the main root surface, shown in the middle, there was hardly any water. The root surface was always surrounded by a black area, indicating a water-deficient site.

Since there are no data on how much water exists at the surface of the root, especially within 1 mm of the root surface, the neutron images shown in Fig. 1.12 were of considerable interest. The evidence of the images that there is hardly any water adjacent to the root surface raises many questions about the water absorption activity of the root itself. How does this phenomenon happen? It is widely known that the composition ratio of the soil matrix, air, and water in cultivated soil for farming is approximately 1:1:1. Too much water in soil is harmful to plant growth.

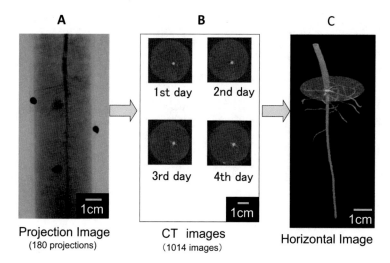

A Projection Image
(180 projections)

B CT images
(1014 images)

C Horizontal Image

1st day   2nd day

3rd day   4th day

Fig. 1.11 Neutron images and root growth. (**a**) projection image of a control soybean sample on the 4th day (1014 × 510 pixels). The whiteness in the figure corresponds to the water amount. Black dots in the image are cadmium standards to adjust the position at CT construction. (**b**) CT images at the same height of the container during 4 days: the large white spot at the center indicates a main root

Roots absorb water from the soil, but why does hardly any water exist immediately adjacent to the root surface?

One of the answers is that growing roots perform movement, called circumnutation (see Chap. 8). Therefore, roots are always searching for favorable conditions in soil to grow, not only for nutrients or water but also for favorable physical conditions of the soil, hard or soft. Because of this movement of the root, the soil around the root tip is put aside during growth, and as a result, a small space is created around the root tip that seems to facilitate growth.

By stacking the dissection images taken every 50 μm, a 3D image of the water and the root imbedded in the soil was constructed. Figure 1.12 shows the CT image of the root imbedded in soil and the 3D image when approximately half of the area of the horizontal CT images was superposed, showing the root.

The whiteness in the image was calibrated well to the water amount. The calibration curve of the whiteness in the figure to water amount is shown in Fig. 1.13.

To visualize the decreased water in the soil more clearly, six reconstructed CT images corresponding to different heights of the container were superposed sequentially on the root image, which was taken from the 3D image of the whole container. These images permitted the comparison of water content changes during growth (Fig. 1.14). Figure 1.14 shows that within a few days, the amount of water in the upper part of the container decreased, corresponding to increased root formation. As shown in the figures, the water holding capacity near the root shifted downward with root development, suggesting movement of the active site in the root. The line profile of the whiteness in the soil along the main root showed that when the side roots were

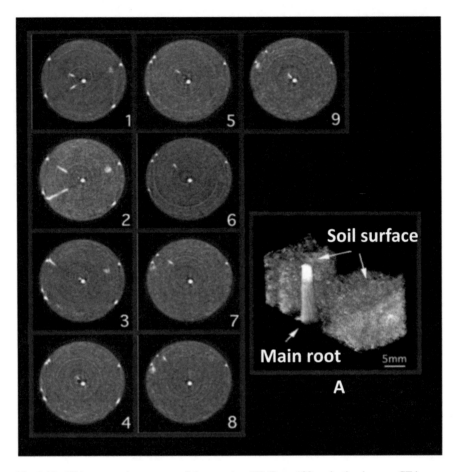

Fig. 1.12 CT images and upper part of the container [2]. From 180 projection images, CT images were constructed. The left figures show the successive CT images constructed every 1 mm from the upper part of the soil to the bottom (1–9). The white spots in the middle show the main root, and the radial white images from the center towards the wall of the container are the side roots that stop at the wall of the container. In most images, there is no water in the neighboring site of the main root, as indicated by the black color, which suggests that the root was absorbing water vapor, not water solution. **A**: spatial image produced by superposing half of the successive 400 CT images, which were taken every 50 μm, at the upper part of the container

about to emerge, the amount of water at those sites decreased (data not shown). When the water profile in the soil was calculated, there was a minimum in the water gradient near the root, approximately 1.0 mm from the root surface. Then, from this point, the water amount sharply increased toward the surface. The root surface was highly wet with more than 0.5 mg/mm^3 of water but not saturated. When Al (10 mM) was applied to the soil, root development and the water holding activity of the root decreased (data not shown).

From the tomographic root images, the position and lengths of the side roots, as well as water movement around them, can be quantified. When the whiteness of the

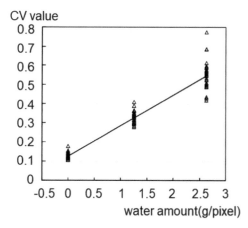

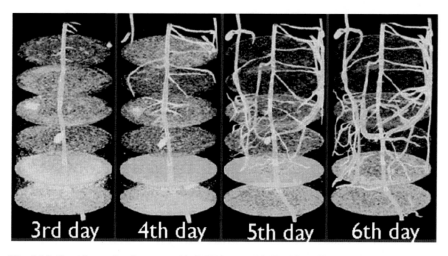

Fig. 1.13 Calibration curve to convert the whiteness value (CV) to water weight (g) [5]. Three standard samples containing 0, 10, and 20% water were measured. Five 10 × 10-pixel areas from 6 CT images were used in each standard sample. Therefore, 30 areas of whiteness were plotted for each sample. The calibration curve was obtained from an average value in each sample

Fig. 1.14 Spatial root development with 6 CT images [6, 7]. After piling up all CT images, the spatial root image was obtained through image analysis. Six CT images constructed at different heights of the container were superposed to the root images. The whiteness in the CT images enabled us to compare the changes in water content. Within a few days, the amount of water in the upper part of the container decreased, which corresponds to an increased formation of roots

3D neutron image within 10 mm of the root was removed, the relative decrease in water amount near the root during root growth could be plotted (Fig. 1.15). The successive neutron images also provided information on side root growth, namely, the length and the emergence sites of the side roots, from which position they initiated their growth.

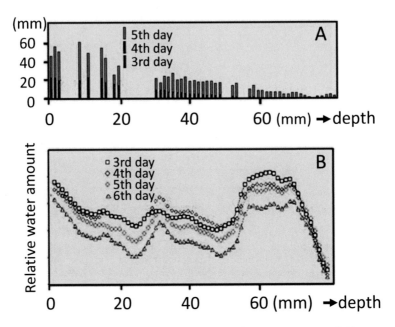

Fig. 1.15 Side root growth and water amount [6, 7]. Quantitative measurement of side root growth according to their position along the main root and water distribution around the main root days after watering the specimen. (**a**) Side root growth and position, (**b**) Water amount along the main root. Depth in mm

There was an obvious increase in side root growth around the upper part of the container (up to 20 mm down from the air/soil interface) compared to side roots located farther down the container. This correlates to a decrease in the water amount found in the soil. However, from this interval of imaging, it was difficult to know whether the decrease in water around the root was due to the horizontal movement of water or to vertical movement along the main root, since water deficiency in some region drives the movement of water toward this site.

Root surfaces and volumes were also calculated from the neutron images. The toxicity of Al ions was analyzed from these calculations since, as previously noted, the presence of aluminum ions is one of the main factors inhibiting plant growth in acidic soil. When 10 mM $AlCl_3$ solution was applied to the soil where a soybean seedling was growing, both the root surface area and root volume decreased [8]. The toxicity of the heavy metal was also visualized by the morphological development of the roots; an example is shown above in Fig. 1.7.

Since neutron imaging has a wide range of effective applications, we tried to apply this technique to other fields. One is fertilizer development, especially in the case of capsule fertilizer, where nutrients seep out gradually during growth (data not shown). Another promising application of this imaging is to study the relationship between root development and yield. It is empirically known that plants with high yield also have well-developed roots. To support this phenomenon with scientific

Fig. 1.16 3D images of roots embedded in soil. *Black dots* are cadmium standards to adjust the position at CT production. (**a**) Soybean root, (**b**) Wheat root

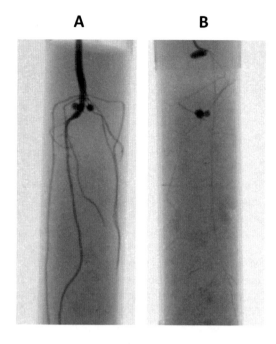

A   B

evidence, further study as to determine how water or fertilizer application promotes growth is desirable, since how to increase crop yields is an important issue all over the world.

Last, as an example of the 3D imaging of roots, spatial images of the soybean and rice roots grown in an aluminum container are presented in Fig. 1.16, where all the CT images were superposed sequentially, from the bottom to the top of the container. Even the fine secondary roots, especially in the case of the rice plant, are clearly shown.

The water profile around the root is not well known since methods of water measurement in small areas of the soil have not been well developed. In field studies, a lysimeter has been used to analyze the moisture balance in the soil, supplying water to the restricted area, and the changes in water amount due to evaporation along with the soil weight were analyzed for calibration. More specifically, water sensors were inserted into soil at certain intervals from the root, and the water amount at the surface of the root was calculated by extrapolating the water profile measured from the sensors, assuming that there was a gradient of water amount toward the surface of the root.

There is no other method comparable to neutron beams to image the water absorption activity of a living root imbedded in soil. Since the roots are the basic tissue that supports plant activity, nondestructive visualization of the morphological development of the roots and of water absorption has a high potential to be applied not only for physiological research on plants but also to the in situ analysis of plants grown in the field.

1.2.3 Water Images of Flowers

When neutron imaging techniques were applied to plant research, first, this method was mainly employed to analyze the aboveground portion of a plant. For example, neutron imaging of a cowpea plant revealed the role of special internode tissue, whose function was to store water (see Chap. 2, Sect. 2.2). Under water-deficient conditions, water is primarily moved from the internode to other tissues. The neutron imaging method was further developed to measure the amount of water actually moving in an internode using ^{15}O-labeled water (see Chap. 2, Sect. 2.3).

One of the reasons neutron imaging was applied to the above-ground part was that it was relatively easy to take the image by simply placing the sample between the outlet of the neutron beam and an X-ray film. As shown in Fig. 1.17, the plant sample was set vertically in front of the cassette in which the X-ray film was placed in vacuum and was irradiated with a neutron beam, by the same method described above for root samples. Figures 1.18, 1.19, and 1.20 are examples of neutron images of flowers: tulip, rose, and convolvulus, respectively. The whiteness in the figure corresponds to the amount of hydrogen in the sample. As mentioned above, at the introduction of neutron imaging, the whiteness could be regarded as an image of the water in a living plant since more than 80% of plant tissue consists of water. Whiter regions indicate the water-rich regions, and the change in whiteness during the treatment indicates the change in the water amount.

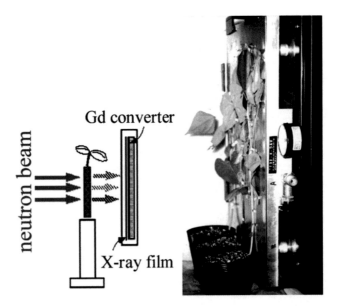

Fig. 1.17 Neutron beam imaging for the aboveground part of the plant. The sample was placed vertically in front of the cassette where the Cd converter and an X-ray film were sealed in vacuum. The neutron beam was irradiated to the sample, and the neutrons after penetrating the sample were converted to β-rays to produce the image on an X-ray film in a cassette

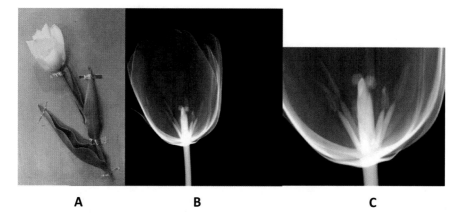

A   **B**   **C**

Fig. 1.18 Water image of a tulip flower by the neutron beam. The degree of penetration of the neutron beam highly depends on the amount of water in the sample. The whiter part indicates higher water content. The neutron beam cannot penetrate the site with high water content; therefore, the exposure of the X-ray film with converted radiation from neutrons behind the plants was low, which resulted in a whiter image. Through calibration, the amount of water in the tissue can be obtained. (**a**) picture of the tulip flower; (**b**) neutron image of the tulip flower; (**c**) magnification of (**b**)

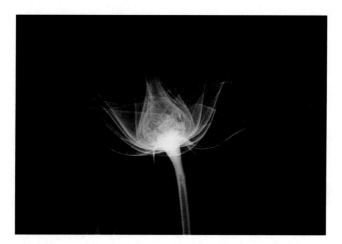

Fig. 1.19 Water image of a rose flower by the neutron beam

As an example of water analysis in a cut flower, a rose image is introduced. An important issue in the cut flower industry, especially for roses, is how to extend the flowering stage. In the case of a rose flower, sometimes the "bent neck" phenomenon occurs during the shipping of the cut flower. Once this phenomenon occurs, the bent neck never returns to the straight position and withers. The bending phenomenon always occurs at the stem, very close to the bottom part of the flower, and was hypothesized to be induced by water deficiency at this part of the stem. To determine from which regions the rose flower loses water, neutron imaging was employed.

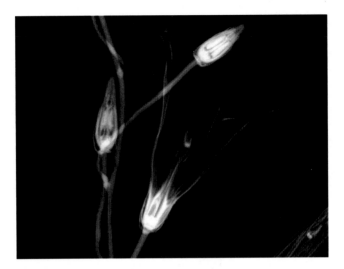

Fig. 1.20 Water image of a convolvulus by the neutron beam

Fig. 1.21 Rose flower images before and after the bent-neck phenomenon. (**a**) Picture of rose flowers; (**b** and **c**) Water image of rose flowers before and after the bent-neck phenomenon. Under dry conditions, the water amount in the internode decreased and could not support the flower part to remain straight, so the flower part was bent

Figure 1.21 is a picture of cut rose flowers and neutron images of these flowers before and after the bent neck phenomenon occurred. To take the neutron image, flowers were fixed on an aluminum board with aluminum tape, as shown in

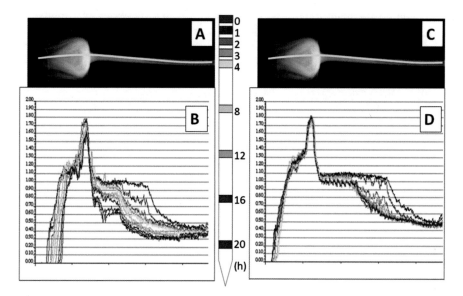

Fig. 1.22   Water amount in a cut flower of rose. (**a**) Water image of a rose flower acquired during the drying process; (**b**) Water image of the rose flower when water was supplied after the drying treatment. The line profile of the water amount was plotted after 0, 1, 2, 3, 4, 6, 12, 16, and 20 h of treatment. Different colors according to time were added to the line profile

Fig. 1.22. When the bent neck phenomenon occurred, the stem had become thinner with less water.

To analyze the water decrease in more detail, neutron images of the flower were taken after 1, 2, 3, 4, 8, 12, 16, and 20 h of the drying process. Before the drying treatment began, 0 h, the water content in the cut flower was very high at the bottom part of the flower and the upper part of the stem, close to the flower. The bending phenomenon occurs at this part of the stem, which initially contained a high amount of water compared to the lower part of the stem. The water content at this site of the stem was maintained well during the first several hours of drying treatment. However, after 8 h, the water in this part gradually decreased.

Next, after 4 h of drying treatment, when the water in the upper part of the stem was not decreased dramatically, water was resupplied. By supplying water again from the bottom part of the stem, the decrease in water at the flower ceased, and water was gradually restored in the stem. However, water was hardly restored at the highest part of the stem close to the flower. The result that water was not reabsorbed at this site in the stem suggested that this site has a special function in maintaining the flowering stage.

That is, when drying conditions began, this part was resistant to decreasing water; however, once water had decreased, water was hardly restored to this part of the stem. This feature might indicate a plant strategy for survival under drying conditions. It seems to require much energy to proceed to the seed ripening stage after the senescence of the flower. Therefore, when the surrounding conditions were not favorable for the plant to develop seeds, such as under low-water availability, the plant could stop the ripening process by discarding the flower part. Bending the higher part of the stem allowed the plant to spend less energy on further development for seed production. However, this example of neutron imaging of cut flowers involves some speculation.

Extension of the life of cut flowers is a key issue in the floral industry, and the amount of water in a flower plays a key role in maintaining its flowering stage. Neutron imaging of cut flowers showed detailed images of water content in flowers such as lilies, morning glories, and carnation. The next is the case of carnation cut flowers, where some attempts were made to extend the flowering stage by supplying modified water. The viscosity of the water was increased by dissolving Xe gas under high pressure, and the prepared water was supplied to a carnation flower to determine whether it helped to prolong the life time of flowering after it was harvested. To acquire the neutron image of the flower, the flower part of the carnation was wrapped softly with aluminum foil, and 180 projection images were taken in the same way as for the root sample. Then, a 3D image of the carnation flower was constructed from the projection images, and the effect of modified water was analyzed by comparing the sagittal images produced. First, the 3D images of a carnation flower demonstrated the importance of water inside the ovule to maintain the flowering stage (Fig. 1.23). The application of water containing dissolved Xe gas to a cut flower helps to control the metabolism of the flower, slowing the deterioration process mediated by an enzymatic reaction (Fig. 1.24).

1.2.4   Water Images of Wood Disks

The green moisture in a wood disk image was shown for the first time by neutron beam irradiation. Image analysis also showed how moisture in wood disks is decreased during the drying process.

There are several features of moisture distribution in the trunk of Sugi (*Cryptomeria japonica*), which is a popular wood utilized in housing materials in Japan. In many cases, there is a so-called white zone at the inner part of sapwood adjacent to the heart wood, recognized by its whiter color. The white zone consists of several annual rings and contains less moisture than surrounding tissues. Green moisture content, especially in heartwood, differs drastically among cultivars or even among individual trees of the same cultivar. It is not known what causes the difference in green moisture content in heartwood, genetics or environmental conditions where it grows. It was reported that there is a reciprocal correlation between the darkness of the color tone and the moisture content in heartwood, where with increasing moisture content, the color becomes darker. Until the tree is cut down, it is not known whether the moisture content of the heartwood is high. Therefore, it is important to distinguish trees with lower or higher moisture content in the heartwood before cutting them down.

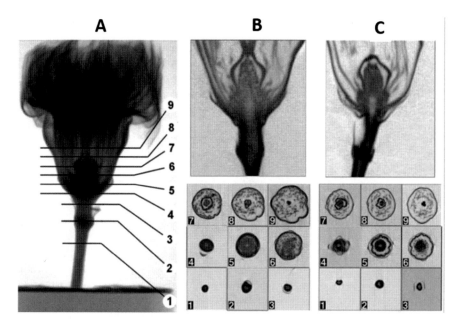

Fig. 1.23 Three-dimensional water image of carnation flowers [9]. Carnation flowers were wrapped with aluminum foil and rotated during neutron beam irradiation similarly to the roots embedded in soil. When all cross-section images constructed at every 50 μm interval at height were obtained, they were superimposed to construct the three-dimensional water image. The lower part of the flower (2 cm) was selected, and transverse sections at each height are shown. (**a**) 3D image of the flower; (**b** and **c**) Cross-section images at nine heights indicated in A before and after the drying treatment, respectively, where the plant was kept under 30° in a phytotron for 2 h without water supply

When the green moisture content in heartwood was high, drying took longer. Since residual moisture after drying lowers the quality of the lumber, eliminating moisture completely from the heartwood is one of the serious problems in kiln drying. To study the decrease in water during the drying process, four cultivars of Sugi, 24-year-old 25-Gou, 25-year-old Honjiro, 29-year-old 1-Gou, and 30-year-old Sanbusugi were cut down in the University Forest in Chiba, Faculty of Agriculture, The Univ. of Tokyo. Approximately 60 cm of the log at breast height was removed by a chain saw, and each end of the log was sealed tightly with vinyl sheets to prevent moisture loss due to evaporation. The next day, the logs were taken into an atomic reactor, JRR-3 M, installed at Japan Atomic Energy Institute, and cut further to obtain wood disks approximately 1 cm in thickness immediately prior to neutron irradiation. The diameters of the disks were 16.6, 8.4, 8.6, and 13.7 cm for 25-Gou, Honjiro, 1-Gou, and Sanbusugi, respectively.

The disks were fixed on an aluminum cassette by aluminum tape where a gadolinium converter (25 μm in thickness) and an X-ray film (Kodak SR) were sealed in vacuum. The cassette with samples was set perpendicular to the neutron beam and irradiated for 19 s. Ten minutes was needed to cool down the sample.

Fig. 1.24 Water images in
the carnation flower when
Xe-gas-dissolved water was
supplied [10]. 2D-neutron
images of a carnation flower
after being supplied with
water containing xenon. The
darker areas indicate water-
rich parts. The xenon in
water maintained a higher
water amount; therefore, the
flowering stage was
maintained longer in the cut
carnation flower

Before water supply

30 min. after water supply

170 min. after water supply

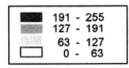

After irradiation, the wood disks were kept in a phytotron at 60 °C with 90%
humidity to reduce the moisture in the disk. During this drying treatment, the wood
disk was periodically removed from the phytotron, and neutron irradiation was
performed in the same manner as described above. Then, the X-ray film was
developed carefully, and the image on the film was transmitted to a computer
through a CCD camera (Hamamatsu Co, 2330).

Figure 1.25 shows the picture and corresponding neutron image of wood disks of
four cultivars. In the neutron images, a whiter color indicated regions where green

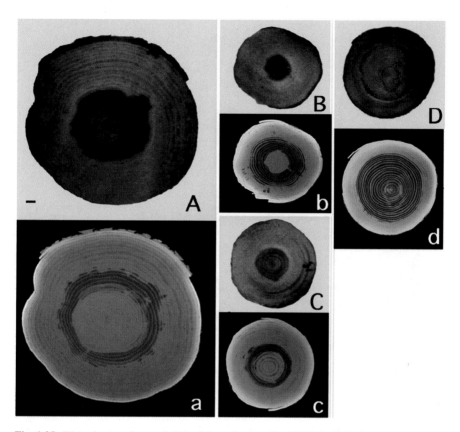

Fig. 1.25 Water image of a wood disk of the cedar tree, Sugi [11]. Optical pictures and neutron images of four cultivars of Sugi (*Cryptomeria japonica*), 25-Gou (**A**, *a*), Honjiro (**B**, *b*), 1-Gou (**C**, *c*), and Sanbusugi (**D**, *d*) before drying. The upper figures for each type (**A–D**) are photographs of the sample disks, and the lower figures (*a–d*) show the corresponding neutron images. The *bar* indicates 1 cm for all figures. The *darkness* in the neutron image corresponds to the degree of moisture deficiency

moisture content was higher. The darkness in the image showed the extremely low moisture content in the intermediate zone between heartwood and sapwood in most of the cases. Within one annual ring, the neutron image showed a regular moisture distribution pattern, both in heartwood and sapwood. Therefore, tracing the moisture-deficient zone that appeared repeatedly throughout the disk enabled us to know the position and number of the annual ring. Among the four cultivars, 25-Gou (a) and Sanbusugi (d) showed the highest and lowest water content in the heartwood, as revealed by the color tone in the neutron images, (A) and (D), respectively. As noted above, it is empirically known that when the color of the heartwood is dark, the moisture content tends to be higher, which was consistent, as shown in Fig. 1.25.

Since the residual moisture in lumber after drying is the largest problem in the utilization of Sugi as a material, two cultivars with different water profiles, as shown

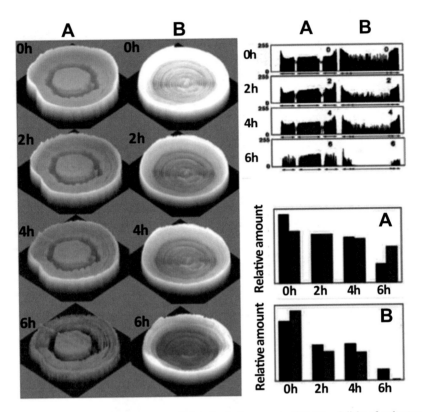

Fig. 1.26 Water image of the cedar tree during the drying process [11]. Wood disks of cedar trees, which were 1 cm thick, were removed every 2 h during the drying process, and neutron images were obtained. The 2D neutron images were converted to 3D images of the wood disk where the height of the image corresponds to the relative moisture content. The *upper-most figure* shows the green moisture pattern, and the downward successive images indicate those after 0, 2, 4, and 6 h of drying treatment. (**a**) and (**b**) correspond to 25-Gou and Sanbusugi, respectively. *Upper right*: Water profile along the line, including the center, across the neutron image of the disk. *Lower right*: relative water amount at the heart wood and sap wood of the area in the line profile of the wood disk at fight high. *Purple region*: heart wood; *Black region*: sap wood

in Fig. 1.25, were chosen, and the changes in water content and profile in the wood disks during the drying process were studied by neutron imaging. The samples were kept in a phytotron at 60 °C with 90% humidity. Every 2 h, the wood disk was removed, and a neutron image was taken. Figure 1.26 shows the neutron images of 25-Gou and Sanbusugi, which, respectively, had the highest and lowest water content in heartwood among the four cultivars, taken during the drying process. The 2D neutron images shown in Fig. 1.25 were converted by a computer to produce 3D images in which whiteness in the image corresponded to the height, so that the higher regions represented water-rich regions. Then, the line profile of each image across the middle was taken out to show the dissection image of water in the disk. The pattern of decreasing moisture on the plane of the disk showed that the moisture of the heartwood tended to remain at a higher level in 25 Gou. The water decrease during the drying process was also clearly indicated when the heartwood area in the dissection image was plotted to show the relative change in the water amount in the disk.

The results can be summarized as follows. In short, when the water content in heartwood was high, the color tone of the heartwood in the neutron image was whiter, and the water amount in the heartwood remained high during drying treatment, whereas when the water content in the heartwood was low, shown by a darker color in the image, the water in heartwood was more easily lost during the drying treatment. The main part of lumber utilized for manufacturing is the heartwood, and water remaining in the wood after manufacturing will induce fractures or cracks during years of usage. However, there is no information about what causes the difference in water amount in heartwood. Indeed, even neighboring trees of the same cultivar showed different water content in the heartwood; the water content was known only after felling the trees.

To obtain additional water images of trees, several kinds of trees grown in the University Forest were felled, and logs 80 cm in length were prepared from a portion at approximately breast height 4 h before neutron imaging was performed. Special care was taken to cover the surface of the log to prevent water loss before preparing the disk for imaging. Then, the wood disks were prepared immediately prior to irradiation and irradiated with a neutron beam, by the same method described above for Sugi. From the images taken, two kinds of trees with interesting water profiles are presented in Fig. 1.26. These images represent wood disks (1 cm in thickness) of metasequoia (*Metasequoia glyptostroboides*) and Japanese cypress (*Chamaecyparis obtusa*), which were 10 and 19 years old, whose sizes were 13.2 and 12.3 cm in diameter, respectively. Similar to the neutron images of Sugi, there was a repeating pattern of white areas in the disk indicating the annual rings. In the case of metasequoia, although there was no color change observable in the picture taken by a camera, the neutron image showed a clear difference in water distribution. The outer several rings contained much higher amounts of water, whereas the water content in the center part was low. The distribution pattern of water within the disk suggested that heartwood formation had already started with no change in the color tone (Fig. 1.27).

In the case of the cypress, white color, indicating a water-rich part, was observed, especially at a few outer annual rings. However, from place to place, the water-rich part penetrated a few annual rings. It was interesting that there was no rigid border for water distribution, i.e., water was not confined within one annual ring but seemed to move across the annual rings. If this image indicated water movement across the annual rings, many questions arose. Since there are many ions dissolved in water, the ions could move with water. The movement of ions indicates information movement associated with water movement. The transfer of information across the annual rings suggests that there is another activity within the heartwood and sapwood, information transfer. There is a method to measure the age of the trees by analyzing ^{14}C activity in the carbon contained at a specific annual ring. Since the decrease in ^{14}C activity correlates well with the number of annual rings actually counted, it seems

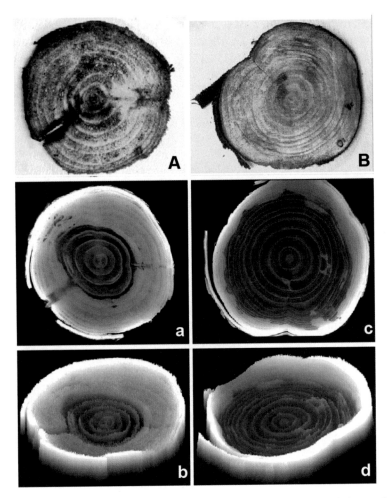

Fig. 1.27 Water image of metasequoia and Japanese cypress. Water images of the wood disks (1 cm in thickness) of metasequoia (*Metasequoia glyptostroboides*) (**A**, *a*, *b*) and Japanese cypress (*Chamaecyparis obtusa*) (**B**, *c*, *d*) are shown. **A** and **B**: Photographs of disks. The 2D neutron images (*a*, *c*) were converted to 3D images (*c*, *d*) by a computer where the height corresponded to the water amount, which is the whiter part in the 2D images

that the carbon structure in the trunk hardly moves, and only water moves in the structure of the carbon network created within the trunk.

We always wondered why most of the tissue in a tree consisted of dead cells. Most living cells exist, particularly in the outer layers of the trunk, and proliferate to enlarge the structure. To maintain the structure of the trees, large amounts of tissues are needed to produce hard trunks. However, when all tissues consist of living cells, they require a large amount of nutrition and energy. The most efficient way to support the structure might be that most of the trees consist of dead cells that do not require energy. However, information must move within the tree, such as to decide when to start the formation of heartwood or to enlarge the heartwood volume.

All the cells in the layer surrounding the heartwood in the sapwood should become dead cells to join the heartwood year by year. It is known that a few percent of the cells in sapwood are alive. When the heartwood increases, how these cells in sapwood, next to the heartwood, become dead cells is not well understood. Are the few living cells in sapwood killed at the border with the heartwood? Although we are apt to focus on the activities of living cells to study plant activity, the role of dead cells might be taken into account.

Important aspects of the physiological activity of a tree could rely on knowing the water and element distribution inside the wood. However, another perspective is to study the element distribution within a tree. We investigated the element distribution within a tree grown in a tropical rainforest, where annual rings were not formed because the weather is relatively constant throughout the year. Through activation analysis of the elements, the distribution of the ion concentration within the disk showed each ion-specific gradient or pattern (data not shown), suggesting that the transmission of each ion might have a specific role in the trunk; therefore, it seemed that there was specific information transfer activity within the tree trunk.

1.2.5 Water Images of Seeds

What kind of image can be taken when seeds are irradiated with neutron beams? Figure 1.28 is a neutron image of corn seeds at an early stage of germination. As shown in the figure, the water distribution within the seed is not uniform, and there is always a water-rich part that will grow as a root to develop toward the outside of the seed.

To visualize how seeds absorb water from outside and how water moves or accumulates in the seed to germinate, water absorption images of the seeds were taken. Five seeds were prepared for neutron irradiation during the water absorption

Fig. 1.28 Water images of corn seed germination

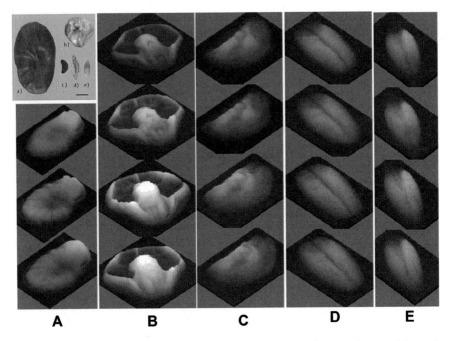

Fig. 1.29 Water absorption process of seeds [12]. Three-dimensional neutron images of the seeds during the water absorption process are shown from top to bottom. Five types of seeds were dipped in water: broad bean (**A**, *a*), corn (**B**, *b*), morning glory (**C**, *c*), wheat (**D**, *d*), and rice (**E**, *e*), and neutron imaging was performed after 1, 2, and 3 h. In the case of the broad bean, the images before and after 1 and 2 h of water treatment are shown. The height in the image represents the water amount, i.e., the whiteness in the image. The standard sample was simultaneously irradiated to normalize the whiteness of the image taken at different stages of water absorption. The distribution of water after absorption was not uniform in all seeds. Water preferentially accumulated at the embryo site in seeds. *Upper left*: photograph of five seeds. The *bar* indicates 5 mm

process: broadbean (*Vicia faba* L.), corn (*Zea mays* L. cv. Kou 504), morning glory (*Ipomoea nil* L. cv. Murasaki), wheat (*Triticum aestivum* L. cv. Minorimugi), and rice (*Oryza sativa* L. cv. Norin 61). The seeds were dipped into water, and after 1, 2, and 3 h, they were removed from the water and wiped well to remove the water remaining at the surface of the seeds. Then, the seeds were fixed on an aluminum cassette where an n/γ converter and an X-ray film were sealed in vacuum. The neutron irradiation of the seeds was performed repeatedly in the same way used for the other samples cited above. To visualize the water distribution within the seeds more clearly, the 2D seed image was converted to a 3D image, where the height indicates whiteness, that is, the amount of water.

As shown in Fig. 1.29, the whiteness in seeds increased with increasing water amount in seeds, but the increase was not uniform. There seemed to be a specific route for water to move within the seeds. In the case of broad bean (a) and corn (b), the water amount was especially high in the embryo. However, in the case of wheat (d) and rice (c), water uptake was found to be higher at the endosperm. Swelling of

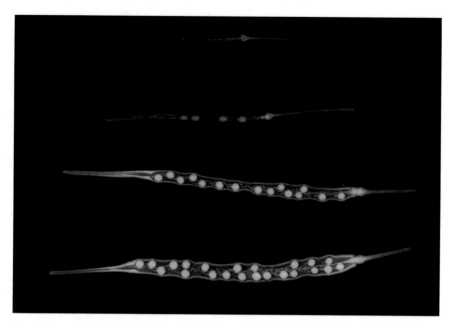

Fig. 1.30   Water images of the rape plant pods [13]. The rape plant pods after 7, 14, 20, and 27 days of flowering from top to bottom

the embryonic root and shoot was clearly observed in the morning glory (c). Since the neutron images showed only the distribution of the water amount and not the route of water movement, it was not known how the water actually moved within the seeds. Several kinds of water channels, aquaporins, are reported in plants; are there many kinds of aquaporins working at different sites in the seeds at different times? Another question regards the subsequent stages of germination. For example, it is not known how the seed skin is triggered and which part of the seed skin opens to allow the root to emerge from the seed. How water moves within the plant could provide some clues to answer such questions.

Several rape plant pods (*Brassica naps* L.) were harvested at different developmental stages of seed formation, and neutron images were taken, since a pod is an important tissue where the optimum conditions for seed development are created. During the ripening of the pod, the water content of the seed reached a maximum at approximately 20–30 days, at an early stage, and then began to decrease, whereas the weight of each seed gradually increased throughout the ripening process because of oil formation. Therefore, there should be a drastic change in chemical transport through the vascular system during the ripening process, including limitation of the water supply. Figure 1.30 shows neutron images of the pods. From the water distribution in the pod, the formation of the pod parenchyma as well as the vascular system could be clearly observed. The formation of the wall in the pod was distinguished at an early stage, and the seeds seemed to emerge from the pod parenchyma. The vascular system was shown to be connected to the shrunken wall. This rape seed pod imaging is an example, but pod visualization is applicable to various developments in agricultural technology, especially the formation of sterile plants, where nondestructive analysis is needed.

1.3 Summary and Further Discussion

The neutron images provide water-specific images of plants that are not obtainable with other means. The CCD camera employed had the highest resolution (approximately 16 μm) of any other CCD camera available. One pixel in the figure corresponded to approximately 16 μm; therefore, we considered the resolution of the image to be approximately 16 μm. We expect that a CCD camera with higher resolution will be developed and that the water movement inside single cells might be able to be analyzed by neutron imaging.

The neutron imaging of roots was introduced. Through neutron imaging, morphological development of the roots imbedded in soil and water movement close to the root could be visualized, and not only 2D images but also the construction of 3D images enabled analysis of the morphological pattern of the roots and the water profile in soil.

Image analysis of the roots imbedded in soil showed that the water absorption activity in the roots gradually shifted downward from the upper to the lower part. Additionally, from the images adjacent to the root, the water absorption activity at the specific site of the root increased before development of the side roots. The water amount near the root had a gradient that increased drastically from 1 mm from the root surface toward the root, but the root surface was still not saturated with water. The space around the root surface was clearly observed when a 3D image of the root was created. In particular, the dissection image of the root imbedded in soil showed a water-deficient area in the vicinity of the main root surface, from the upper to the lower part. This suggests that there is hardly any contact between water solution and the root surface, which might indicate that the roots are absorbing water vapor rather than water solution. The root tip is always searching for a favorable location and creating space for the root tip to grow in the soil by circumnutation. In the case of a rice plant, one cycle of root tip rotation was found to take 50 min (see Chap. 8). Therefore, there is always a space in the vicinity of the root surface. If the roots absorb water vapor, the absorption of nutritional elements, including metals, from the roots should be drastically different from that of roots growing in water culture. Are the roots absorbing metal vapor, too?

The orientation of root development is another interesting topic, but it was not discussed much here. How does the root decide the direction of the growth? When VA mycorrhizal fungus was placed in a thin box and neutron images were taken, one of the side roots of a soybean plant developed linearly toward the fungus, which was more than 5 cm from the root (data not shown). Whether the hyphae induced the root pattern or not was not discernible.

The analysis of the morphological development of the root enabled us to evaluate the condition of the soil, especially in the development of soil fertilizer, soil conditioners or supplying devices of water. However, there was no reproducibility in the formation of the root pattern, since it is impossible to prepare exactly the same soil conditions for root growth. Therefore, as one solution to evaluate the soil condition based on root development, we measured the root length after formation of the line profile of the root as an indicator to compare growing conditions.

In the case of flower imaging, the water profile within the stem, bulb or pod could be visualized. Water movement derived from neutron imaging suggests tissue-specific functions for water, such as in stems. The bending neck phenomenon suggested another function of the stem. In the case of a rose plant, all of the stem close to the flower consisted of living cells, whereas in other plants, pith was formed where dead cells were packed. During a water-deficient period, living cell activity might cease and thereby prevent water movement toward flower parts. When the pith consisted of dead cells, it could be assumed that water might move easily by capillary phenomena. It was interesting to note that the function of dead cells might also be taken into account.

Many approaches are attempted to extend flowering; one of them is to change the viscosity of water, such as by dissolving an inert gas, such as Xe gas. When water containing dissolved Xe gas was supplied to a carnation flower, the flower took longer to become senescent, suggesting that the water loss due to transpiration was decreased. How water movement is controlled was proposed to provide clues regarding plant activities.

The visualization of green moisture in a wood disk was the first example. There were some other applications of neutron imaging in wood samples, such as detection of wood discoloration in a canker fungus-inoculated Sugi [2] or to measure the decay resistance of chemically modified wood [14], etc. The visualization of drying process of Sugi was presented in this chapter. Because Sugi is a popular type of lumber in Japan for building houses or furniture, and only one species exists. Even among the same cultivar of Sugi, the water content in the heartwood drastically varies. It is not known what causes the differences in moisture content in the heartwood. Water distribution within the wood disk showed a water gradient pattern within each annual ring so that the number of the water peaks was the same as the number of annual rings. In the case of Sugi, there was a difference in water content between heartwood and sapwood, and there were always a few water-deficient annual rings at the boundary between heartwood and sapwood, called white rings. However, the size of the white ring increases as the heartwood and sapwood grow. Then, how is water moved across this white ring toward the heartwood while maintaining the white ring? With the development of the tree, the size of the heartwood increases, and white rings are still formed between the heart and sap-wood. That is, the white ring grows and maintains a low amount of water, whereas the amount of water in the heartwood or sapwood is maintained during growth. Is there any system to increase the size of this white ring containing a low amount of water?

There are many possibilities for the application of neutron imaging techniques to study plant samples. The neutron images acquired revealed new aspects of plant activity but raised many questions at the same time. Some of the questions raised were about root activities, but another interesting question was about the function or role of dead cells. The dead cells support the plant activities of living cells, which could be estimated through water images; however, how the dead cells are used efficiently is not known. Another question is how plants control water movement. The movement of water is a kind of engine that induces not only water absorption but also growth. It seemed that water movement was not derived simply from diffusion or osmotic pressure.

Though neutron imaging showed static water images of the plant samples, when successive static images were taken, it was possible to estimate the plant activity based on water movement, since the movement of the plants is rather slow.

Then, in the next section, the real movement of water in a living plant is presented utilizing radioisotopes.

Bibliography

1. Nakanishi TM, Matsubayashi M (1997) Nondestructive water imaging by neutron beam analysis in living plants. J Plant Physiol 151:442–445
2. Nakanishi TM, Okuni Y, Furukawa J, Tanoi K, Yokota H, Ikeue N, Matsubayashi M, Uchida H, Tsuji A (2003) Water movement in a plant sample by neutron beam analysis as well as positron emission tracer imaging system. J Radioanal Nucl Chem 255(1):149–153
3. Saito K, Nakanishi TM, Matsubayashi M, Meshitsuka G (1997) Development of new lignin derivatives as soil conditioning agents by radical sulfonation and alkaline-oxygen treatment. Mokuzai Gakkaishi 43(8):669–677. (Japanese)
4. Nakanishi TM, Matsumoto S, Kobayashi H (1993) Water hydrology by neutron radiography when water absorbing polymer was added to the soil. Radioisotopes 42:30–34
5. Okuni Y, Furukawa J, Matsubayashi M, Nakanishi TM (2001) Water accumulation in the vicinity of a soybean root imbedded in soil revealed by neutron beam. Anal Sci 17 (supplement):1499–1501
6. Furukawa J, Nakanishi TM, Matsubayashi M (2001) Water uptake activity in soybean root revealed by neutron beam imaging. Nondestr Test Evaluat 16:335–343
7. Nakanishi TM (2009) Neutron imaging applied to plant physiology. In: Anderson IS, McGreevy RL, Bilheux HZ (eds) Neutron imaging and applications, a reference for the imaging community. Springer, Cham, pp 305–317
8. Habu N, Nagasawa Y, Samejima M, Nakanishi TM (2006) The effect of substituent distribution on the decay resistance of chemically modified wood. Int Biodeterior Biodegradation 57:57–62
9. Nakanishi TM, Furukawa J, Matsubayashi M (1999) A preliminary study of CT imaging of water in carnation flower. Nucl Instrum Methods Phys Res A424:136–141
10. Matsushima U, Ooshita T, Nakanishi TM, Matsubayashi M, Seo Y, Kawagoe Y (2000) Effect of non-polar- gas on water in cut carnation flowers. Journal of the Japanese Society of Agricultural. Machinery 62(5):70–78
11. Nakanishi TM, Okano T, Karakama I, Ishihara T, Matsubayashi M (1998) Three dimensional imaging of moisture in wood disk by neutron beam during drying process. Holzforschung 52:673–676
12. Nakanishi TM, Matsubayashi M (1997) Water imaging of seeds by neutron beam. Bioimages 5 (2):45–48

13. Nakanishi TM, Inanaga S, Kobayashi H (1991) Non-destructive analysis of rape plant pod by neutron radiography. Radioisotopes 40:126–128
14. Yamada T, Oki Y, Yamato M, Komatsu M, Kusumoto D, Suzuki K, Nakanishi TM (2005) Detection of wood discoloration in a canker fungus-inoculated Japanese cedar by neutron radiography. J Radioanal Nucl Chem 264:329–332

Chapter 2
Real-Time Water Movement in a Plant

Keywords Real-time water measurement · Water moving speed · Plant · Soybean ·
^{18}F · ^{18}F-water · ^{15}O · ^{15}O-water · Water circulation · Water leakage from xylem ·
Water moving orientation · Cowpea · Draught tolerant

2.1 RI-labeled water

To analyze the water absorption from the root to the aboveground part of the plant, in more detail, the basic questions are how water is absorbed by the roots, how water is transported upward from the roots, and how water interacts with the surrounding tissue of the xylem during transport. To study these questions, radioisotope tracer work is indispensable both for imaging and for the measurement of trace amounts of water.

What kind of RI can be used to label water? Since water molecules consist of two elements, H and O, there are limited radioactive nuclides applicable to label water, ^{3}H and ^{15}O. The method is to supply radioactive water to the plant and measure the radiation from the RI emitted by the plant. However, there is an important condition required to perform nondestructive measurement of the radioactive water supplied to a living plant. That is, the radiation emitted from the radioactive nuclides within the plant should be high enough to penetrate the plant tissue and can be detected from outside the plant by the counter prepared.

In the case of hydrogen, the only candidate radioactive nuclide for tracer work is tritium, ^{3}H. However, the β-ray energy of ^{3}H is too low (max. energy: 18.6 keV) to penetrate the plant tissue and be detected. Therefore, the β-rays from ^{3}H cannot be measured from outside the plant tissue. Therefore, ^{3}H cannot be used for the nondestructive real-time imaging or analysis of water in a living plant. Tritium can be employed only for destructive experiments. To analyze the ^{3}H incorporated in the plant, the tissue must be removed from the plant and exposed to an IP to acquire the image, and the tissue should be digested by chemicals to measure the radiation amount by a liquid ion counter or other detector.

Then, the most promising candidate radioisotope to label water molecules is ^{15}O, which is a positron emitter; however, there is always a problem of positron escape

T. M. Nakanishi, *Novel Plant Imaging and Analysis*,
https://doi.org/10.1007/978-981-33-4992-6_2

when using a positron emitter for imaging. Positron escape makes it impossible to calculate and compare the water amounts among the different tissues in the image. This problem will be described in more detail in the next section. Another problem in employing ^{15}O is that the half-life is extremely short, only 2.04 min. Therefore, the experiment can last only approximately 20 min, and data must be calibrated according to the half-life.

There is another candidate for tracing water, ^{18}F. One of the representative methods of producing of ^{18}F is by employing the nuclear reaction $^{16}O(\alpha, pn)^{18}F$. Thus, when water is irradiated with an α (He) beam, trace amounts of ^{18}F are produced in the water as a carrier-free nuclide. The carrier-free radioisotope is known to move with the surrounding molecule. Since the half-life of ^{18}F is relatively long (110 min) compared to that of ^{15}O (2.04 min), this nuclide was employed to measure water movement. Both nuclides, ^{15}O and ^{18}F, are positron emitters, and when the labeled water was supplied to the plant, the radiation from these radioisotopes in the plant could be detected from outside the plant. The radiation showed where the labeled water was, and the radiation counts could, with care, be converted to the water amounts actually present or moving within the plant tissue.

2.1.1   Positron Emitters

Before presenting the application of positron emitters, ^{15}O and ^{18}F, for tracing water movement, the features of positron emitters are presented briefly. Several kinds of positron emitting nuclides that can be applied for tracer work are produced by using an accelerator (Table 2.1).

In the medical field, the utilization of positron emitters for nondestructive imaging of the human body is called positron emission tomography (PET). Various kinds of positron emitters are supplied to humans to diagnose lesions or tumors. However,

Table 2.1 Production of positron-emitting nuclides

Nuclide	Half-life	e^+- energy	Production	Target	E_{th}
^{11}C	20.4 (m)	0.961 MeV	^{14}N (p, α) ^{11}C	N_2	3.1 MeV
			^{10}B (d, n) ^{11}C	B_2O_3	3.0
^{13}N	9.97 (m)	1.20	^{16}O (p, α) ^{13}N	CO_2, H_2O	5.5
			^{12}C (d, n) ^{13}N	CO_2	0.3
			^{13}C (p, n) ^{13}N	$^{13}CO_2$	3.2
^{15}O	2.04 (m)	1.73	^{13}N (p, n) ^{15}O	$^{15}N_2$	3.7
			^{14}N (d, n) ^{115}O	N_2	
^{18}F	110 (m)	0.634	^{18}O (p, n) ^{18}F	H_2 ^{18}O, $^{18}O_2$	2.6
			^{20}Ne (d, α) ^{18}F	Ne_2	
			^{16}O (α, pn) ^{18}F	H_2O	
^{48}V	15.9 (d)	0.697	Ti (p, xn) ^{48}V	Ti	
			^{45}Sc (α, n) ^{48}V	Sc	

these positron emitters have not been utilized well for plant research because of certain problems.

The radioactive nuclides called positron emitters emit positrons (β^+) during the decay process. The emitted β^+ soon combine with electrons (β^-) and produce two identical γ-rays that were emitted in 180 degree opposite directions to each other at the same time, with identical energy, 0.511 keV. This phenomenon is called annihilation. In the medical field, positron emitters are widely used for two main reasons. One is that the useful positron emitters have relatively short half-lives so that after application, the radioactive nuclide will soon decay, keeping the radiation effect on humans low. The other reason is that the same detection system can be used for different positron emitters because the measurement is based on the same energy of the two γ-rays produced by annihilation, 0.511 keV. However, these measurement systems cannot be applied to plant studies since positrons easily escape from thin plant tissue. The details are described in the next section.

2.1.2   Positron Escape Phenomenon

As noted above, when a positron emitter is employed to visualize the distribution of the nuclide in an intact plant, there is a problem called "positron escape." Since the positron emitted from the radionuclide has relatively high energy, the positron can move a considerable distance before undergoing the annihilation process to produce γ-rays. This means that the position of the positron emitting nuclide itself in the sample could be different from that of the γ-ray produced. The detection system focuses on the sample and measures the γ-rays from the sample, regardless of the positron escape phenomenon. The positron detector consists of a pair of γ-ray detectors for simultaneous counting for the two identical γ-rays emitted in opposite directions; therefore, the detector focuses on the sample, which is set between the two counters. In the case of a thin tissue, most of the positrons emitted from the positron emitting nuclide in the sample, could escape from the tissue and produce γ-rays in the air outside the sample. We cannot measure the γ-rays in the air, and as a result, the apparent amount of positron emitters in the image is very low.

Therefore, there is always a problem in the real-time imaging of the positron emitter in a plant because this positron escape phenomenon is dependent on the thickness of the plant tissues.

When the plant tissue is thin, like a leaf, a significant part of the positrons escape from the tissue before annihilation occurs, and as a result, the apparent amount of the nuclide at the leaf in the image is very small. As a result, positron images of a plant always show a clear image of the internode with a fading image of leaves. Therefore, to perform radioactivity counting of a positron emitter in the image of a plant over time, the counting area should be fixed at the same position in the image, and the counting at this site cannot be compared to that at other sites. Therefore, internode counting is preferable to counting the leaves.

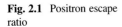 **Fig. 2.1** Positron escape ratio

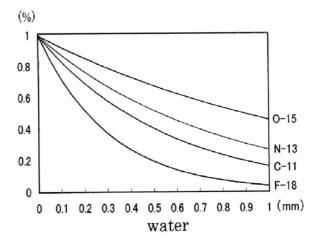

The distance between the positron emitter site in the sample and the γ-ray production site causes the degradation of resolution. The positron escape ratio in water is shown in Fig. 2.1. Considering that more than 80% of the plant tissue consists of water, the distance between the positron emitting nuclide in tissue and the site of the γ-ray detected site after the annihilation process is sometimes on the order of mm.

The positron escape phenomenon also occurs in the human body in PET, and there is also a distance between the tumor, where the positron emitter accumulates, and the site where γ-rays are produced and detected. Therefore, the resolution of PET cannot be less than the order of mm. However, currently, with the assistance of image processing techniques, PET images can be modified to produce a smooth distribution of the nuclide in the tissue. However, the positron imaging system used for PET in humans cannot be applied to the plant samples, primarily because of the morphological features of the plant. A plant is largely a collection of thin sections and is very different from the human body, which is just a mass.

2.1.3   Production of RI-Labeled Water

Two kinds of radioisotope-labeled water are candidates to analyze the real-time water movement in a plant: ^{18}F-water and ^{15}O-water. Since the half-lives of ^{18}F and ^{15}O are short, 110 and 2.0 m, respectively, the nuclides have to be prepared just before usage and finish the experiment before decaying out the radioisotopes, i.e., while radioactivity can be detected. The preparation method of ^{18}F and ^{15}O is shown in Fig. 2.2.

To produce ^{18}F, 6 g of ice prepared in a Ti vial was placed in an aluminum container. Then, this container was irradiated with a He beam (50 MeV) for 40 min by an AVF cyclotron installed at JAEA (Japan Atomic Energy Agency) to produce

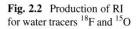

Fig. 2.2 Production of RI for water tracers ^{18}F and ^{15}O

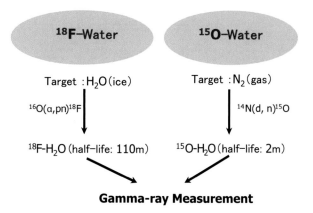

^{18}F in water by the nuclear reaction ^{16}O(α, pn)^{18}F. After irradiation, the water was passed through a cation exchange resin to remove impurities, such as ^{48}V, produced in the Ti vial.

To obtain ^{15}O, N$_2$ gas was irradiated with a 12 MeV deuteron (d) beam to produce ^{15}O by the nuclear reaction ^{14}N(d, n)^{15}O, using an accelerator at QST (National Institutes for Quantum and Radiological Science and Technology, Japan). By irradiation, ^{15}O-labeled gas was produced, which was then introduced to distilled water, where a Pt catalyst at 120 °C was used to catalyze an exchange reaction of ^{16}O with ^{15}O to produce ^{15}O-labeled water.

As noted above, tritium, ^{3}H, is another candidate to label water as a tracer. However, since the β-ray energy (18.6 keV) from ^{3}H is too low to penetrate the plant tissue and cannot be measured from outside the plant tissue, ^{3}H was not used for real-time imaging to analyze water in a living plant. Tritium was employed only for destructive experiments to acquire 2D images of water distribution by an imaging plate (IP) or for radioactivity counting by a β-ray counter after chemical digestion.

The following sections present the results we obtained by employing ^{18}F$^-$ and ^{15}O-labeled water to trace real-time water movement in a living plant.

2.2 ^{18}F-Water (Half-Life is 110 min): Cowpea, What Is Drought Tolerance?

2.2.1   System of ^{18}F-Water Imaging

When water is irradiated with a helium beam, ^{18}F is produced from oxygen, a constituent of water molecules, by the ^{6}O(α, pn) ^{18}F reaction, as described above. Considering the half-life of ^{18}F (110 min), the tracer solution applied to the plant was adjusted to 10 MBq/ml, rather high radioactivity as a tracer.

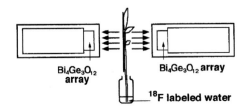

Fig. 2.3 Schematic illustration of ^{18}F-water imaging [1] ^{18}F-water was supplied to the plant, and the γ-rays produced by positrons emitted from ^{18}F were detected by a pair of BGO arrays

The γ-rays emitted from ^{18}F in the plant were measured by a pair of $Bi_4Ge_3O_{12}$ (BGO) scintillation detectors (Fig. 2.3). Each detector consisted of an array of small BGO detectors with a detection area of 2×2 mm. The aboveground part of the plant was fixed on a nylon mesh board with tape and set vertically between the two detectors. The detectors faced each other, and only simultaneous γ-ray counts by both detectors were recorded since the γ-rays were emitted at the same time by an annihilation phenomenon. The spatial resolution of the image obtained by this positron imaging system was estimated to be approximately 2.4 mm, which was the highest resolution obtained using the array of small $Bi_4Ge_3O_{12}$ scintillation detectors. The target area was 5×6 cm, and the real-time radioactivity of water in the target area was monitored with a computer. First, a cowpea plant was chosen to trace water movement within a plant using [18]F.

2.2.2 Cowpea

Cowpea (*Vigna unguliculata* Walp) was employed to study water absorption because it is widely grown in the semiarid regions of India and Africa for its high resistance to drought conditions. This plant is considered one of the most drought-resistant species among pulse crops. It was reported that under drought conditions, cowpea leaves were able to maintain comparatively high water potential and hence relatively high photosynthetic activity. As a mechanism of drought resistance in cowpea plants, it was suggested that the stem had a water storing function; however, such water storing tissue in the stem has not yet been identified, possibly due to technical difficulty.

2.2.3 Neutron Imaging of Cowpea

First, a neutron image of the cowpea plant was taken to determine the water distribution within the plant. As mentioned in Chap. 1, the neutron image shows a water-specific image. The neutron image acquired is shown in Fig. 2.4.

Fig. 2.4 Water image of a cowpea plant by neutron beam irradiation [2] The whiter part of the figure corresponds to the site with more water. From the extent of whiteness in the figure, cowpea was found to have a water-rich tissue in the primary leaf internode

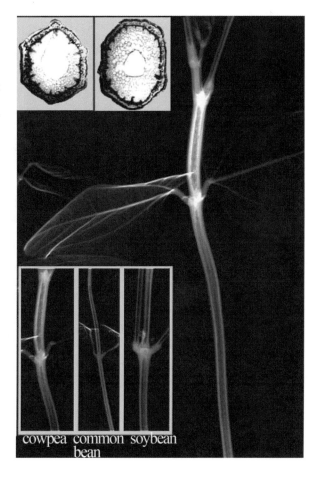

cowpea common soybean
bean

As described in Chap. 1, the whiteness in the figure corresponds to the water amount. The whiter part of the figure corresponds to areas with more water. From the extent of whiteness in the figure, cowpea was found to have a water-rich tissue in the primary leaf internode, suggesting that this part of the plant functions as the potential water storage tissue.

Neutron images of the primary leaf internodes of soybean and common bean were taken, and only cowpea was found to have water-rich tissue in the primary leaf internode (Fig. 2.4). By comparing the cross section of the primary internode with that of the cotyledonous internode, it was found that parechymatous tissue in the primary leaf node was well developed for water storage, although a similar number of cells were counted at both internodes.

To determine the distribution of water in the internode more specifically, the whiteness of the image was dosimetrically scanned across the diameter of the primary leaf internode as well as the cotyledonous internode, and it was found that only in cowpea plants was the water content particularly high at any point in the

Fig. 2.5 Line profile of the internode in the water image [2]. Line profile of the internode in cowpea (**a**), common bean (**b**), and soybean (**c**) plants. The vertical axis shows the magnitude of whiteness, which indicates the water content. The horizontal axis shows the diametric distance across the internode. C: center; S: surface of the internode; Solid line: line profile across the internode between the primary leaf and the first trifoliate; Dashed line: internode between cotyledon and primary leaf. For the cowpea plant, the primary leaf internode contained a very high amount of water

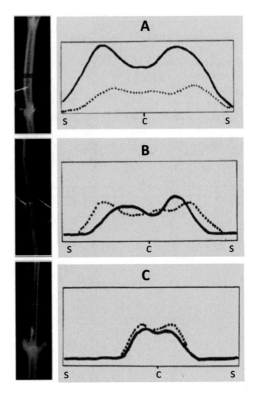

primary leaf internode. When the internodes of cowpea, common bean and soybean were compared, no morphological differences were observed, but neutron radiography clearly showed the difference in water content (Fig. 2.5).

Although a water storage function of the parenchymatous tissue was reported for plants growing in the desert (6,9), this cowpea plant imaging was the first direct observation of the water storage tissue.

2.2.4 ^{18}F-Water Uptake of Cowpea

After the physiological properties of the cowpea plant were analyzed, ^{18}F-water was supplied to the plant for only 1 min from the lower part of the stem, where the roots were cut off. A picture of the plant target is shown in Fig. 2.6. Then, distilled water was supplied instead of ^{18}F-water. Since the target area was 5 × 6 cm, the area to image water uptake included the primary leaf internode and the first trifoliate leaf. The real-time radioactivity observed in the water accumulation image of the target area was monitored every 1 min. Figure 2.7 is an example of a water image in a cowpea plant. The lower figure is a photograph of the plant target, and the upper figure is an example of an ^{18}F-water image based on a 1 min accumulation of counts.

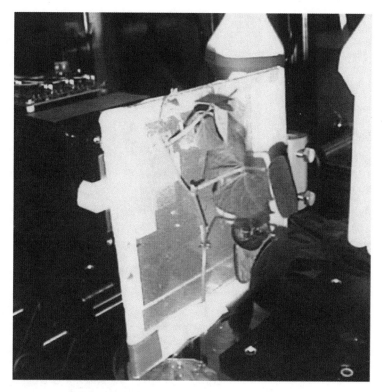

Fig. 2.6 Example of setting the plant sample for imaging. The plant sample for imaging was set on a mesh and vertically between two detectors

The successive water accumulation images of the target were integrated every minute until 60 min, and the results showed that water accumulated first in the primary leaf internode and gradually moved up to the first trifoliate leaf (Fig. 2.8).

To understand the behavior of water uptake with time more clearly, the radioactivity counts of the two sites, the primary leaf internode and the first trifoliate leaf, were extracted from the series of the images in Fig. 2.8 and plotted until 30 min (Fig. 2.9). When ^{18}F-water was supplied, the water absorption speed was very rapid at first, and then, after approximately 10 m, the increase in radioactivity plateaued. This tendency of ^{18}F-water absorption was similar between the cowpea and common bean. The amount of water moved up to the primary leaf internode was approximately 2–3 times higher than that in the first trifoliate leaf in both plants. This difference in ^{18}F counting between the internode and the leaf is caused by the positron escape phenomenon. As explained earlier, when the thickness of the tissue is approximately 0.2 mm, as in a leaf, approximately half of the positrons escape from the leaf tissue; therefore, radiation counting comes to approximately half that in thick tissue, such as an internode (Fig. 2.1). However, following the radioactivity at the same site made it possible to compare the change in counts with time at that site, since the escape ratio of positrons based on the tissue thickness did not change.

Fig. 2.7 Cowpea plant and an example ^{18}F-mediated water image [1]. Two cowpea plants were selected for imaging, including the primary internode (S) and the first trifoliate leaves (L). (**a**) Photograph of the imaging plants. An imaging area of the plants is indicated in square. (**b**) Example ^{18}F-water image of the target. The size of the target area was 50×60 mm. The ^{18}F signals at S and L were plotted for the absorption curve of the cowpea plant

After 1 h of drying treatment, the ^{18}F-water absorption in common bean was drastically changed from that in cowpea. Figure 2.9 shows the ^{18}F-water uptake behavior of both plants before and after drying. Although the water transport behavior of the cowpea and the common bean was similar before the treatment, the water uptake activity of the common bean plant was drastically decreased after drying, whereas cowpea was shown to maintain high water uptake activity.

When leaf photosynthesis (LPS) activity before and after water depletion treatment in cowpea and common bean was compared, the LPS activity of the cowpea plant was maintained at a high level compared to that of the common bean (data not

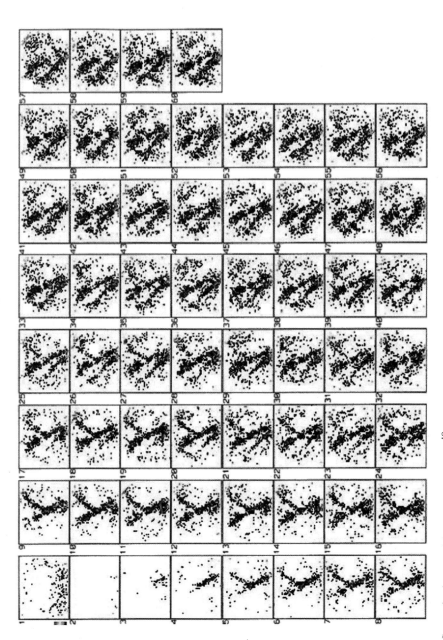

Fig. 2.8 Successive images of the cowpea plant after ^{18}F was supplied [1]. Successive images of the cowpea plant were acquired every 1 min of accumulation until 60 min after ^{18}F-water was supplied from the lower part of the stem. The high ^{18}F signal in the first image was due to the background γ-rays, since the shielding was incomplete only at this setting time

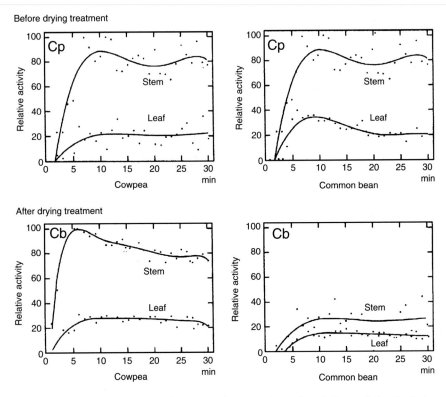

Fig. 2.9 Water absorption curves of cowpea and common bean plants before and after the drying treatment [1]. Successive water absorption curves of cowpea plants (Cp) and common bean plants (Cb) before and after the drying treatment. The counting regions of stem and leaf of the cowpea plant are shown in Fig. 2.7 as S and L, respectively. For the common bean, identical sites to those of the cowpea were selected

shown). The water image of the cowpea plant by neutron beam showed an extremely high amount of water in the primary leaf internode compared to the other internodes. The image suggested that in a cowpea plant, when water was depleted, the water in the storage tissue seemed to play an important role in maintaining both photosynthesis and water uptake activity, thereby maintaining tolerance of water-depleting conditions.

2.2.5   What Is Drought Tolerance?

Finally, an unexpected aspect of the water uptake activity of drought-tolerant cowpea is presented. After establishing the drought tolerance of cowpea compared to the common bean, the next question was whether there is any difference in the degree or features of drought tolerance among the cowpea. To study differences in drought tolerance within cowpea, experiments were performed on drought-resistant and drought-sensitive cowpea selected in Africa from approximately 2000 naturally grown cowpea plants. They were not artificially created by crossing, gene engineering or other methods but acquired drought-tolerant and drought-sensitive properties through evolution.

Before starting the ^{18}F-water supply and the measurement, the nature of the tolerance was estimated. It was taken as natural that water absorption ability is the key factor in whether a plant is drought tolerant or drought sensitive. The drought-sensitive plant must have low-water absorption activity so that under drought conditions, it loses the ability to absorb water. Therefore, it is also natural to think that to make the drought-sensitive plant tolerant, its water absorption ability should be enhanced. The study then examines how the high water uptake activity in tolerant plants is regulated or maintained and determines which gene is responsible or would be effective to introduce. This approach is, of course, widely accepted an important strategy to pursue.

Using the naturally selected drought-tolerant and drought-sensitive cowpea, the fixed internode site was selected, and the water accumulating at this site was measured, considering the positron escape phenomenon. When ^{18}F-water was supplied to the naturally selected plants, to our great surprise, the water-absorbing activity in the sensitive plant was much higher than that in the tolerant plant under normal conditions (Fig. 2.10). That is, the water-sensitive plant always required

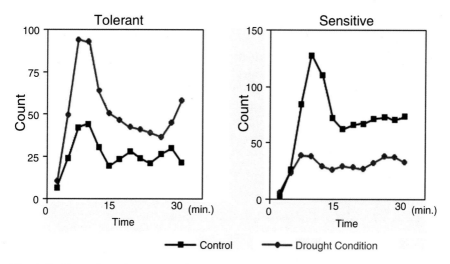

Fig. 2.10 Water absorption curves of drought-tolerant and drought-sensitive cowpea plants. Drought-tolerant and drought-sensitive cowpea plants were selected in Africa from approximately 2000 cowpea plants naturally grown in the field. Drought-tolerant cowpea commonly absorbed less water than sensitive cowpea, which was expected to be the reverse phenomenon. However, under the drought treatment, the amount of absorbed water greatly increased. The sensitive sample absorbed more water than the tolerant sample but could not absorb water after the drying treatment. Vertical axis: relative amount

more water than the tolerant plant. However, the drought-tolerant cowpea absorbed less water than the drought-sensitive cowpea. However, after treatment with drought conditions, drought-sensitive plants could not absorb high amounts of water under normal conditions, as shown. However, the tolerant plant showed increased water absorption activity, and the amount of water absorbed was much higher than usual.

This change in water absorption activity was unexpected, especially because it contradicted our assumptions about the mechanism of water absorption in the laboratory. Especially for the drought-tolerant plant, the mechanisms of maintaining lower water absorption than that of drought-sensitive plants under normal conditions and of triggering enhanced water absorption activity are not known. However, it seemed as if the tolerant cowpea was always using less energy, perhaps reserving energy for water absorption to prepare for the emergency of a drought.

2.3 ^{15}O-Water (Half-Life Only 2 min): Water Circulation Within an Internode

As presented above, ^{18}F-water measurements could trace water movement. However, there always remains a question of whether the behavior of F^- ions is exactly the same as that of water molecules. The number of ^{18}F atoms produced in 1 g of water (3.3×10^{22}) targeted by a cyclotron was 3×10^7. Therefore, because of the small ionic radius of F^- ions, the trace amount of ^{18}F could be expected to move with the overwhelming amount of water molecules in a plant. However, to eliminate this uncertainty, the preferable radioactive nuclide to label water is ^{15}O, with an extremely short half-life of 2 min. In the next step, ^{15}O-labeled water was employed to study water transport in a soybean plant. As the first step to perform quantitative analysis of water movement with ^{15}O-water, an imaging plate (IP) was employed to acquire successive static images with time. Then, a real-time measuring system for ^{15}O-water movement was designed that enabled quantitative analysis. Since ^{15}O is a positron emitter, like ^{18}F, the positron escape phenomenon was consistently taken into account.

2.3.1 ^{15}O-Water Image in the Internode

^{15}O-water was produced by the nuclear reaction ^{14}N(d, n)^{15}O, as mentioned in 2-1-3. The ^{15}O-water was prepared with a radioactivity concentration of 2 GBq/10 ml and supplied to the plant. Then, ^{15}O-water imaging by an IP was performed. The plant was fixed on a board, and an IP was placed as close as possible to the board. Since the half-life of ^{15}O is extremely short, only 2 min, the ^{15}O-water exposure time was set to 1 min to acquire the image. Exchanging the IP with a new one, another 1 min exposure was performed to acquire the subsequent image. By comparing the two

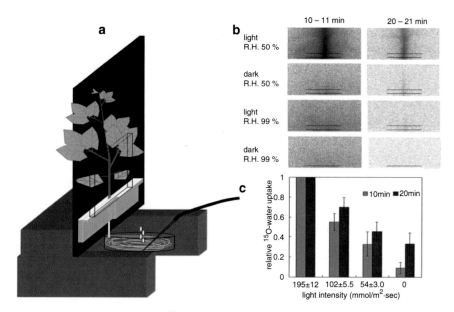

Fig. 2.11 Usage of an IP to image ^{15}O-water absorbed in a soybean plant [3]. A soybean plant was fixed on the board, and ^{15}O-water was supplied from the root. An IP was placed on the board for 1 min to obtain a ^{15}O-water image of the plant. (**a**) An IP was placed near the lower part of the aboveground part of the plant. To acquire the successive image, the IP was replaced with a new one after 10 and 20 min of the ^{15}O-water supply from the root. (**b**) Internode image of the plant under different humidity and light conditions. R.H.: relative humidity; (**c**) uptake amount of ^{15}O-water in the internode under different light intensities

images obtained from the IPs, it was able to show the change in the ^{15}O-water profile, which indicated water movement. As an example, ^{15}O-water images in a soybean plant (*Glycine max* cv. Tsurunoko) are presented. Figure 2.11a is a schematic illustration of acquiring the RI image of the living plant.

After ^{15}O-water was supplied from the root, the water image at the lower part of the internode was acquired, since the internode was a preferable tissue for quantitative analysis because of the lower occurrence of positron escape. The IP was placed on the board from 10 min to 11 min and from 20 min to 21 min after the ^{15}O-water supply began. The images of ^{15}O-water in the internode are shown in Fig. 2.11b. Since the half-life of ^{15}O is extremely short at 2 min, the image of the internode soon disappeared. As shown in the figure, the darkness of the image across the internode could be converted to the amount of ^{15}O-water absorbed. With changes in humidity and the light intensity used to irradiate the plant, the amount of water taken up by the plant changed drastically. When the relative amount of water taken up to the internode during 10–11 min and 20–21 min under 50% humidity and under the highest light intensity was set as 1, the amount of water taken up by the plant was clearly dependent on the light intensity (Fig. 2.11c).

2.3.2   Water Movement Is Different from that of Cd Ions

Another example of ^{15}O-water imaging with an IP showed that the uptake of ions and that of water itself are different. In the case of Cd ion, the amount absorbed in a soybean plant was different under different pH conditions. However, there was no information on whether the amount of water absorbed was different from that of Cd ion. When the water amount absorbed is higher than that of Cd ion, it suggests that the plant is diluting the Cd concentration at absorption or during transport.

To study this plant activity, ^{109}Cd (0.325 kBq/ml) solution was supplied to an 8-day seedling of a soybean plant for 2 h under different pH conditions. Then, the distribution ^{109}Cd in the plant was imaged by an IP. The accumulation of ^{109}Cd in the aboveground part of the soybean plant was much higher under lower pH (4.5) culture conditions than under a higher pH (6.5), close to neutral. In contrast, when ^{15}O-water was supplied and the ^{15}O distribution was imaged by an IP, the amount of ^{15}O-water absorbed under lower pH (4.5) conditions was lower than that under higher pH (6.5) conditions (Figs. 2.12 and 2.13). Because of the extremely short half-life of ^{15}O, the profile of newly absorbed water was obtained for only a short time (5 min) after ^{15}O-water was supplied. The opposite profiles of Cd and water in the aboveground part of the plant suggested that the heavy element Cd dissolved in water did not move together with the water flow in the plant. Under low pH conditions, the amount of ^{109}Cd in the aboveground part was much higher, whereas

Fig. 2.12 ^{109}Cd absorption image of soybean plants under different pH conditions [4]. ^{109}Cd distribution 4 days after ^{109}Cd was supplied from the roots under different pH conditions

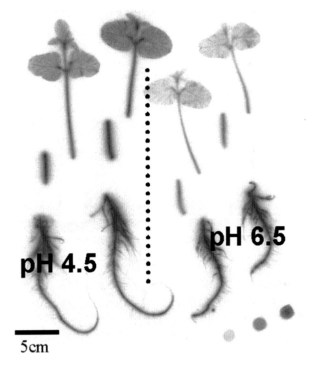

pH 4.5   pH 6.5

5cm

Fig. 2.13 ^{15}O-water absorption image of soybean plants under different pH conditions. ^{15}O-water distribution after 5 min of ^{15}O-water supply under different pH conditions. Although the half-life of ^{15}O is extremely short, it showed the difference in amount of water in the aboveground part of the plant between pH 6.5 and pH 4.5. This water profile image was opposite to that of the ^{109}Cd distribution in Fig. 2.12

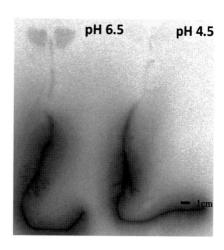

the amount of ^{15}O-water transferred to the aboveground part was less than that at pH 6.5, suggesting that the concentration of Cd in the aboveground part was higher under acidic conditions, which might increase the toxicity of Cd in the plant. Another interpretation was that the water movement toward the aboveground part was suppressed and the upward Cd movement increased under acidic conditions. Although the IP image enables quantitative analysis, it is a static image; therefore, there was a limit to the information provided by the image.

2.3.3 Real-Time Water Movement in a Plant

The ^{18}F imaging method using an array of $Bi_4Ge_3O_{12}$ (BGO) detectors was difficult to apply to ^{15}O-water imaging to analyze the real-time movement of ^{15}O because of the positron escape phenomenon, which presented a serious problem in the image analysis. With that imaging method, it was not possible to calculate how much water was actually moving from one tissue site to the others in the image because of the different thicknesses among the tissues. Therefore, to trace the radioactivity change with time, the analysis site had to be fixed at one site in the image. Considering the features of positron emitters, the measuring system had to be redesigned to measure the amount of ^{15}O-water moving in real time within a plant. When the imaging target was fixed to a small site of the plant (for example, see Fig. 2.10), it was possible to trace the change in radioactivity with time. Since ^{15}O has an extremely short half-life and decays out rapidly, the counting efficiency of the γ-ray detector was required to be as high as possible, which means that a larger crystal scintillator was needed than that used for ^{18}F. The detection area of the BOG detector was increased from 2×2 mm to 10×10 mm, and to maintain positional information, a small part of the internode, 1 cm in length, was selected. Large detectors were set adjacent to either side of the target to reduce the effect of positron escape from the sample on the

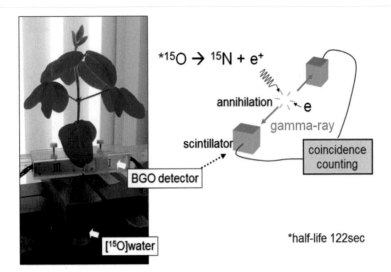

Fig. 2.14 Schematic illustration of ^{15}O counting of an internode when ^{15}O-water was supplied to the soybean. A pair of γ-ray detectors was adjusted to the position for the measurement, i.e., 2 cm above the cotyledon. Counting was performed when two identical γ-rays, which were produced from the annihilation by the positrons emitted from ^{15}O, were simultaneously counted by both BGO detectors. The sample plants are approximately 25 cm in height. After the application of ^{15}O-water, the ^{15}O-water amount accumulated at 1 cm of the internode was counted. The half-life of ^{15}O is 122 s

detector. With this system, the volume of ^{15}O-water in 1 cm of the internode could be measured quantitatively and noninvasively. After ^{15}O-water was supplied to the plant, the radioactivity from ^{15}O at the targeted 1 cm internode was measured by a pair of BGO detectors, and the amount of water accumulating in this tissue was calculated.

Figure 2.14 shows the sample at measurement and a schematic illustration of the radiation counting. When ^{15}O decays, the nuclide is changed to ^{15}N, and at the same time, a positron e$^+$ is emitted. The positron soon reacts with an electron, e, and is converted to two identical γ-rays emitted 180 degrees apart, which is called annihilation. The pair of BGO detectors was used to count these γ-rays from annihilation, and the detector was set as close as possible to sandwich the stem (Fig. 2.15).

2.3.4 Design of ^{15}O-Water Measuring System

The system to measure ^{15}O-water accumulation at the internode of the plant was designed as shown in Fig. 2.15. Figure 2.16 shows a schematic illustration of a BGO detector (crystal size 10 × 10 × 20 mm, Photosensor Modules: Hamamatsu

coincidence measurement

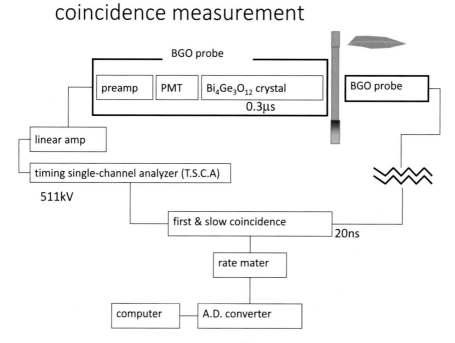

Fig. 2.15 Diagram of the measurement system for ^{15}O-water. A long cable (approximately 4 m) was used to connect the gamma-ray detector with the coincidence circuit. The vial and detectors were shielded by lead blocks. The illustrated parts, except for the coincidence circuit, were set in the growth chamber. PMT: photomultiplier tube

Photonics, Co., Japan) and a picture of the detector placed adjacent to the plant internode. A BGO crystal was coupled to photomultipliers fixed in an aluminum frame.

A pair of BGO probes was installed face to face across the internode, which was 2 cm above the cotyledon of the plant. The signals that entered the detection window were amplified by a linear amplifier (704-4B Oken, Co., Japan), and then discriminated by a Timing S.C.A. (706-2B: Oken, Co.) according to the γ-ray energy of an annihilation, 511 keV. Then, the signals were converted to counts through a coincidence circuit (Fast and Slow Coincidence 708-1B, Oken, Ratemeter S-2293B, Oken, Co.) and recorded in a computer (OptiPlex GX1: Dell. Co. Kawasaki). The timing of coincidence and the data export interval were set at 110 ns and 1 s, respectively. The background noise was recorded to <0.01 cps by the coincidence circuit and shielding made of lead blocks.

Since the radioactivity of ^{15}O is reduced by half every 120 s, the radioactivity counting of ^{15}O–water was able to continue for approximately 1000 s. The counts were accumulated every 30, and the detection limit was estimated to be 0.11 kBq from the least squares fitting curve of observed counts (Fig. 2.17). The counting

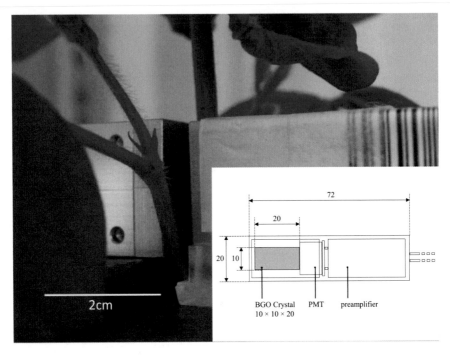

Fig. 2.16 BGO detector. (Photograph of the plant target with a pair of BGO detectors. The detector consisted of a BGO crystal, a PMT, and a preamplifier. The BGO crystal was coupled to photomultipliers fixed in an aluminum frame. The dimensions are given in mm)

efficiency of the γ-ray detector was calculated to be 0.120% by comparing the counts of the BGO probes with that of the γ-ray counts by a counter for the plant tissue supplied with $H_2{}^{15}O$ (200 MBq/ml^{-1}). Table 2.2 summarizes the performance of the measuring system using the BGO detector. This performance of the detector was also confirmed by preparing a phantom of the stem consisting of a silicon tube containing ^{15}O-water gel close to the surface, mimicking the soybean stem with xylem cells (Fig. 2.18).

Figure 2.19 shows an example of ^{15}O-water counting in a soybean stem when ^{15}O-water was supplied. As shown in Fig. 2.19a, the counts from the internode increase at first and then decrease rapidly because of the short half-life of ^{15}O. Therefore, the ^{15}O-water absorption curve was always calibrated with half-life decay, and the curve in Fig. 2.19a was converted to that in Fig. 2.19b.

The measurement system was set in a phytotron, which maintained the conditions throughout the experiment at 27 °C and 50% humidity, and all experiments were performed during the light phase.

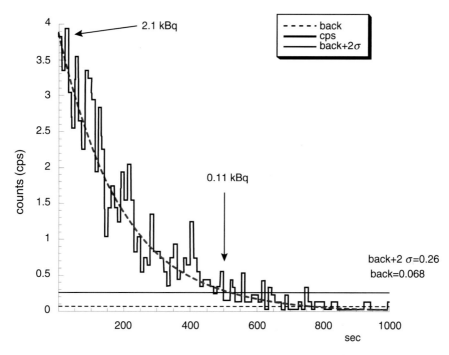

Fig. 2.17 Detection limit of the measurement system. The radioactivity of ¹⁵O-water was counted for 1000 sec, and the counts were accumulated every 30 s. Broken line: background level; solid line: background level + 2σ; broken curve: least-squares fitting exponential curve of the observed counts

Table 2.2 Performance of the measuring system using the BGO detector

Linearity (cps)	0.3–100 cps
Counting efficiency (%)	0.12 ($n = 10$)%
Background (cps)	0.68 ± 0.095 cps
Detection limit (Bq)	0.11 kBq
Efficiency = (¹⁵O activity in a 1 cm stem (Bq)/cps (s⁻¹) × 100%	
Efficiency of calibration tube was 0.14%	

2.3.5 ¹⁵O-Water Absorption Curve

The prepared plant sample was a soybean plant 20 days after germination with second expanded trifoliate leaves. The plant height was approximately 25 cm. After eliminating the cotyledon and excising the root 8 cm below the cotyledon, the bottom part of the stem was placed in a vial. Then, ¹⁵O-water (2 GBq/10 ml) was supplied to the vessel. The vessel and the detectors were shielded by lead blocks to reduce background counts. After ¹⁵O-water was supplied to the plant, the amount of ¹⁵O-water in the 1 cm internode increased linearly at first, and then the slope gradually decreased. Since the half-life of ¹⁵O is only 2 min, measurement could

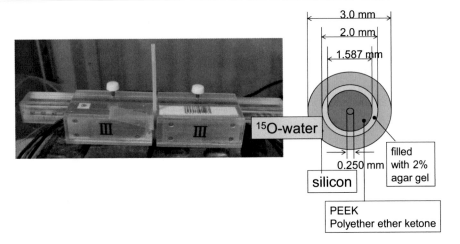

Fig. 2.18 Phantom of the prepared stem. To calibrate the γ-ray counting to the water amount, in addition to the actual gamma-ray counting by cutting the internode, the phantom of the internode was prepared, and calibration was confirmed

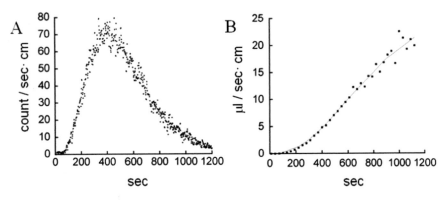

Fig. 2.19 Calibration of the ^{15}O count decay. (**a**) Raw counting data measured by the system. The data were obtained every 10 s; (**b**) ^{15}O-water uptake curve calibrated with the half-life decay and counting efficiency. The data were calculated every 30 s

be performed until approximately 1000 s. The calibrated ^{15}O-water amount plotted in the absorption curve is shown in Fig. 2.20. Surprisingly, the amount of ^{15}O-water measured at the 1 cm internode exceeded the xylem volume (1.9 μL) in the internode within only one minute.

The capacity of the xylem vessels was calculated from the measurement of their transverse sectional area under microscopy $\{1.9 \pm 0.3 \times 10^{-7}\ \text{m}^2$ (total cross section area) $\times\ 1 \times 10^{-2}\ \text{m}$ (length of measuring part of the internode)$\}$. From the linear part of the slope of the absorption curve, the increasing amount of ^{15}O-water per 1 cm of the internode was calculated as $5.2 \pm 0.5 \times 10^{-2}\ \mu\text{L/s}$.

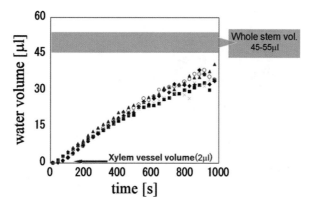

Fig. 2.20 ^{15}O-water uptake curve at 1 cm stem of a soybean plant [5]. ^{15}O-water was supplied to a soybean plant, and the amount of the water in 1 cm of the stem, which was 2 cm above the cotyledon, was measured. Since the half-life of ^{15}O is extremely short (2 min), the absorption measurement could be perform until approximately 1000 s. Within a few minutes, the amount of ^{15}O-water in the stem exceeded the vessel volume in the 1-cm stem (2 µL) and increased to approximately 45–55 mL, which is close to the entire volume of the targeted 1-cm stem, after 1000 s. The results indicate that a large amount of water leaked from the xylem. Each symbol represents the data of individual plant samples ($N = 5$)

During the measurement, the amount of ^{15}O-water continued to increase and occupied a volume of 40 µL after approximately 15 min, which was more than 20 times higher than the vessel capacity and close to the whole volume of the targeted stem, 1 cm in length (45–55 µL).

This ^{15}O-water absorption curve indicated a very interesting result: a tremendous amount of water was always leaking out from the xylem vessel, which had been regarded as a mere pipe to transport water to the surrounding tissues. It was also suggested that the water was leaked out not longitudinally but horizontally. However, continuous lateral water movement must be associated with longitudinal transport, which was derived from transpiration by the plant.

2.3.6   Route of Water Flow Leaked from Xylem

From the measurement of the newly absorbed water movement and accumulation in the internode, it was found that a large amount of water was always leaking out from xylem vessels (Fig. 2.21). The next question was the route of the water after leaking out from the xylem. There are four possible routes for the flow of water leaked from xylem.

1. Flow out from the stem surface.
2. Flow into the phloem vessel.
3. Move upwards in other tissue besides the xylem vessel.

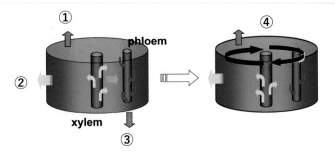

Fig. 2.21 Schematic illustration of four possible routes for water movement after the escape from the xylem vessels: ① water flows upwards through the tissue other than xylem vessels; ② water flows towards the stem surface and evaporates; ③ water flows into the phloem tubes and is transported downward; ④ water re-enters the xylem vessels

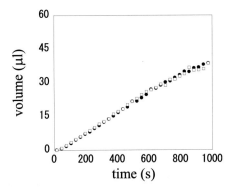

Fig. 2.22 Effect of the Vaseline treatment on the lateral water movement [5]. Water evaporation from the stem surface was inhibited by a Vaseline cover on the surface. The vertical axis indicates the volume in the targeted stem occupied by ^{15}O-water. ^{15}O-water was applied at $t = 0$. The filled circles and open squares quantify the volume of ^{15}O-wate before and after the Vaseline application, respectively

4. Return to the xylem vessel.

1. Flow out from the stem surface.

When water moves toward the surface of the internode and evaporates there, it could provide a driving force for water to move within the internode. To determine whether the water leaked out from the xylem vessel was lost from the internode surface, the surface of the internode was covered with Vaseline, and the ^{15}O-water uptake curve was measured. There was no change in the water absorption curve before and after covering the internode with Vaseline (Fig. 2.22). The results indicated that the large amount of water leaking out from the xylem vessel was not lost from the internode surface, which means that most of the leaked water was not moving toward the surface of the internode.

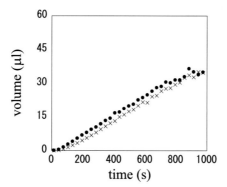

Fig. 2.23 Water uptake curve when cambium was removed [5]. Phloem flow was knocked out by removing the tissue outside of the cambium at 3.0–3.5 cm and 1.5–2.0 cm above and below the measuring position, respectively. The vertical axis indicates the volume in the 1-cm stem occupied by ^{15}O-water. ^{15}O-water was applied at $t = 0$. The filled circles and crosses indicate the volume of ^{15}O-water before and after removing the tissue outside of the cambium, respectively

2. Flow into the phloem vessel.

   Since there are some reports that there is an interaction between xylem and phloem vessels, the outside of the cambium was removed, and the water uptake curve was measured. Figure 2.23 shows that there was no change in the water uptake curve after removing the phloem vessels, indicating that water leaked from the xylem vessels was not moving toward the phloem vessels.

3. Move upwards by other tissue besides the xylem vessel.

   Dissection of the internode of the stem showed clearly that the inside of the xylem vessel was empty, and the other tissues were filled with tissue cells. Therefore, it can easily be estimated that it might be natural for water to move upward within the xylem vessel than through other tissue where resistance significantly decreases the movement speed of the water.

   If the leaked water is moving upward through the tissue other than the xylem vessel, there must be a difference in the velocity of the water movement from that in the xylem. To determine the water velocity within the xylem, the volume of the xylem in the internode was measured. Since the inside of the xylem vessel is empty, when there is no change in the whole volume of the xylem vessel with height, it can be concluded that there is no change in the water velocity within the xylem vessel. When the internodes of the stem, 2 and 6 cm above the cotyledon, were sliced and the area of the xylem vessels was measured under a microscope, the area of the xylem vessels at different heights was almost the same, 1.91 ± 0.32 and 1.86 ± 0.31 ($\times 10^{-7}$ m^2), respectively. This result indicated that the same volume of water was constantly transferred upward within xylem vessels, regardless of the height in the internode, suggesting that the speed of upward water movement in the xylem vessels is the same throughout the internode. The next question concerns the volume of the tissue surrounding the xylem tissue containing the water. As shown in the absorption curve, the ratio of the water volume in the xylem to that in the other tissue, flowing out from the xylem, was approximately 1:20, indicating that the leaked water volume was much higher than that remaining in the xylem. Under the microscope, the area of the xylem vessel and the other tissue was measured, and it was found that the ratio of xylem vessel area to that of the other tissue was approximately 1:20, which could be converted to volume. That means that approximately 20 times more water was contained in the 20 times larger volume of the tissue other than xylem.

Since the ratios of the amount of water and the tissue volume in xylem and the other tissue were approximately the same, the upward velocity of water movement in each tissue must be the same when the leaked water was moving through other tissues. However, as noted above, it cannot be the same because of the structure of xylem and the other tissue.

There is another reason to discount this model, which is the water velocity in the internode shown in Fig. 2.25, where the water velocity in xylem cells remained the same (see next section). In conclusion, this possibility was ruled out as an explanation of the behavior of leaked out water, although the evidence for its rejection is indirect.

4. Return to the xylem vessel

Of the four possibilities for the route of water movement after leaking out from xylem vessels, 1 to 3 were rejected, as discussed above. Therefore, No. 4 remained to explain this phenomenon. That is, most of the water leaking out from the xylem vessels must re-enter the xylem vessels. According to this scenario, the decrease in the gradient of the ^{15}O-water absorption curve along with the height of the measuring position of the internode has to be interpreted as a decrease in the ^{15}O concentration in xylem vessels rather than a decrease in water escape.

2.3.7   Water Flow in the Internode

In the previous section, it was found that freshly absorbed water was circulating within the stem, leaking out from the xylem and returning to the xylem by exchanging water already present in the stem with the newly absorbed water. To determine the exchange behavior of newly absorbed water in xylem with that in the surrounding tissue, ^{15}O-water was supplied for 5 min as a pulse, and then nonradioactive water was supplied to the soybean plant. As shown in Fig. 2.24, ^{15}O-water linearly increased for approximately 300 s and then slowly decreased, indicating that the water from xylem was smoothly exchanged with the water already present in the internode. This ^{15}O-water movement suggested that the ^{15}O-water already spread out horizontally at the internode in 5 min was pushed horizontally toward the xylem by newly absorbed nonradioactive ^{16}O-water, acting as a returning flow, and then moved upward. As a result, after its initial increase, the ^{15}O activity gradually disappeared from the fixed measuring site of the internode.

Another question is about the vertical water flow in the internode. To measure the water movement speed at different heights of the internode, three pairs of BGO

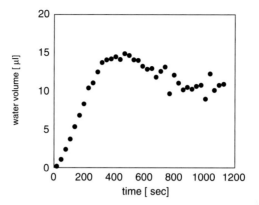

Fig. 2.24 Exchange of ^{15}O-water with stable water and ^{16}O-water. Measurement of ^{15}O at 2 cm above the cotyledon. After ^{15}O-water was supplied for 5 min, stable water (^{16}O-water) was supplied. After the water was changed from ^{15}O-water to ^{16}O-water, the ^{15}O-water amount at the internode continued to increase for approximately 200 s and subsequently gradually decreased, which suggests that water exchange via xylem vessels smoothly occurred

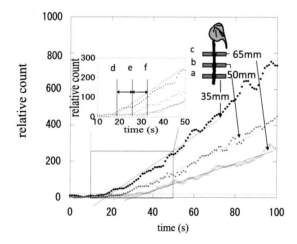

Fig. 2.25 Flow rate of ^{15}O-water in the internode [5]. The amount of ^{15}O-water absorption curve was measured at three positions in the internode: 35, 50, and 65 mm above the cotyledon (a, b, and c, respectively). The intersection of the curve at the *x*-axis after elimination of the noise shows the time when ^{15}O-water was first detected, as indicated by d, e, and f. From the time difference between d and e and between e and f, the flow velocity of water was constant in the internode and calculated to be approximately 4 mm/s

detectors were prepared to measure the water uptake curve. Figure 2.25 shows the ^{15}O-water uptake curve measured at different positions of the internode, 35, 50, and 65 mm above the cotyledon (a, b, and c). From the difference in the time when the detectors first detected the ^{15}O-water, d, e, and f, the water transfer speed in the internode could be calculated. As shown in the figure, the time needed for water to move 15 mm, that is, from a to b and from b to c, was shown to remain the same, indicating that the water movement speed in the internode remained constant (4 mm/s).

Assume that the xylem vessels can be regarded as mere pipes with many surface pores and that water is transported through them. When water escapes from the pipe and does not return to it, the amount of water transported should decrease with height along the pipe. As shown in Fig. 2.25, the flow velocity of ^{15}O-water through the xylem vessel remained almost constant. This result also supported that the water leaked from xylem vessels was returning to the xylem again.

Finally, another experiment was introduced, which simply measured the respiration volume of water from the plant. The evaporation of water was measured by weighing the whole plant in a phytotron, and the evaporation speed was 0.91 ± 0.13 μL/s, including leaves and internodes. As described above, the ^{15}O-water uptake experiment showed that the speed of water leaking from the xylem vessel was 0.052 μL/s/cm (Fig. 2.20). Since the whole internode length was approximately 16.5 cm, the total leaked volume of water from the whole internode was approximately 0.86 μL/s, which was close to the amount of water lost by respiration measured by the weighing experiment, suggesting that the water volume leaking from xylem vessels corresponds to the decrease in water resulting from respiration.

2.3.8   Verification of Water Returning Process to Xylem Using ^{3}H-Water

Another remaining question concerns the horizontal movement of water within the internode. To verify the process of water leaking from and returning to xylem tissue, ^{3}H-water was used instead of ^{15}O-water. Since the β-ray energy emitted from ^{3}H is very low, 18.6 keV, this radiation cannot be detected from outside the plant when ^{3}H-water is supplied to the plant. Therefore, several plants were prepared, and ^{3}H-water was supplied for 5 s to the plants as a pulse. Then, periodically, the internode at the same position of the ^{15}O-water measurement was cut out, and the ^{3}H-water distribution image was acquired by an IP.

Figure 2.26 shows the image of the ^{3}H-water distribution in the internode section after 0, 10, 20, 60, and 120 s of ^{3}H-water supply, with the corresponding picture of the dissection image under a microscope. High accumulation of ^{3}H-water was observed in xylem vessels as early as 5 s of application (time 0 s in Fig. 2.26). After 20 s, ^{3}H-water was spread throughout the transection of the internode. This indicated a rapid inward movement of ^{3}H-water, independent of transpiration. Then, the ^{3}H-water amount decreased, and the intensity of the ^{3}H-water image decreased greatly, indicating that the leaked water returned to the xylem again and moved upwards. Water diffusion is another candidate for water movement. When the diffusion coefficient was 2.4×10^{-9} (m^2/s), the diffusion distance of ^{3}H-water for

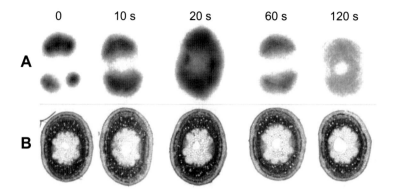

Fig. 2.26 ^{3}H-water uptake shown in the stem sections [5]. To confirm the horizontal leak of water using ^{15}O-water, ^{3}H-water was similarly supplied for 5 s ($t = 0$). Since the β-ray from ^{3}H is too low to detect from outside of the plant, imaging could not be nondestructively performed. The stem was harvested and sliced each time after the ^{3}H-water supply, and the microradiograph of the sliced section was acquired by IP. (**a**) Image of the ^{3}H-water distribution in the internode section, 2 cm above the cotyledon, from the end of the ^{3}H-water supply (0 s) to 120 s. (**b**) Microscopic images corresponding to the upper images. ^{3}H-water leaked from xylem was spread throughout the area of the section after 20 s and gradually returned to xylem tissue

10 cm of the stem could be less than 2 mm after 10 min. Therefore, it was suggested that most of the ^{3}H-water must have been transported longitudinally through the xylem, followed by horizontal diffusion at the point of the measurement.

In short, the results showed that the leaked water spread out horizontally in a short time, flushing away the water already present in the internode and then returning to the xylem vessel again moving upward within the internode.

2.3.9   Summary of Water Circulation Within the Internode

Using both ^{15}O-labeled and ^{3}H-labeled water, the water movement in the internode showed that there was an intense and ongoing lateral water exchange between the xylem vessel and the surrounding tissues along the upward pathway. Therefore, the ^{15}O-water uptake amount decreased because of dilution of the ^{15}O-water in xylem tissue with the nonradioactive water flushed out and returned to the xylem vessel during the ^{15}O-water absorption process. Simulation showed that approximately half of the water already present in the stem was replaced by freshly absorbed water within approximately 20 min (data not shown).

The results of water movement in a soybean plant can be summarized as follows.

1. A tremendous amount of water was always horizontally leaking from xylem tissue.
2. Water that leaked from the xylem tissue then flushed away the water already present in the internode, re-entered the xylem tissue and moved upward, showing water circulation in the stem.
3. The velocity of upward water movement remained constant.
4. In a simulation, within 20 min, half of the water already present in the internode was estimated to exchange with the freshly absorbed water.

2.4 Summary and Further Discussion

Radioisotope-labeled water provided water absorption movement in a living plant. As a representative study, ^{18}F-water uptake in a cowpea plant and ^{15}O-water uptake in a soybean plant are presented.

In the case of a cowpea plant, a special internode was identified, and the water movement under water-deficient treatment suggested that this tissue functioned as water storage tissue to enable heat tolerance. However, it was noted that the water uptake behavior of drought-tolerant and drought-sensitive cowpea naturally produced in the field of Africa showed water-absorbing activity that could not be estimated before. The water-absorbing activity of drought-tolerant cowpea naturally produced in the field was lower than that of the drought-sensitive cowpea, but the water absorption activity of the drought-tolerant cowpea increased greatly under drought conditions, whereas the sensitive cowpea showed the opposite effect. The plant strategy for tolerating water depletion suggested that enhancing water absorption activity is not the way to achieve drought tolerance; instead, growing with limited water requirements is necessary. However, why the water absorption activity of the tolerant plant was suddenly enhanced upon exposure to low-water conditions is still not known.

Although ^{18}F is produced as a carrier-free ion by irradiating water with a ^{4}He beam, whether ^{18}F^{-} has the same behavior as that of water itself remains uncertain. Therefore, ^{15}O was produced and used to trace water movement in a soybean plant. Since the half-life of ^{15}O is extremely short, 2 min, the experiment was performed before the ^{15}O decayed out, within approximately 20 min.

Using ^{15}O-water, it was found that a tremendous amount of water was constantly leaking out from the xylem and returning to the xylem again after flushing out the water that was already present in the xylem vessel. This circulation of water in an internode was observed for the first time. The renewal time of the water already existing in the internode was rather fast. It took approximately 20 min to replace half of the water already present in the internode with newly absorbed water, according to a simulation. Since an internode is packed with cells, and each cell has a cell membrane and organelles inside, it is totally unknown how the water could enter these different micro-organelles easily and flush out the water already present. The

A   **B**

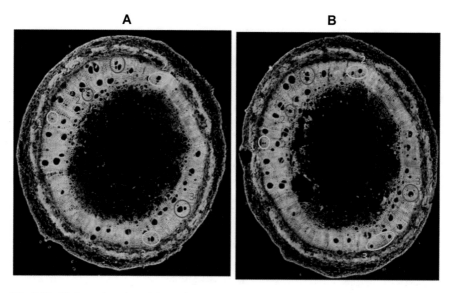

Fig. 2.27 Horizontal section of the soybean stem under a microscope. To confirm the location and size of xylem tissues, sliced sections at different heights are shown. The locations of A and B are 2 mm apart from each other. The location and size of each xylem tissue are different. Numbers 1 and 2 3 in the same color are the same xylem tissue

only known water movement channel is aquaporin. Are there networks of different kinds of aquaporins in the internode?

The xylem vessel is apt to be regarded as a mere pipe to transfer water from the root to the upper tissues. However, the vessel itself has a microscopic structure that differs from place to place. Figure 2.27 shows two cross sections of an internode of a soybean plant, only 2 mm apart. Within 2 mm of distance, some of the vessels disappear, and some new ones appear. This means that the structure of the xylem vessels forms a complicated network throughout the internode, with microscopic changes in the morphological shape and size of the vessels or connection sites with height. However, it was interesting to observe that not only the area of vessels but also the areas of the sieve tube, pith and xylem remained in the same range even when they were 4 cm apart in the internode (Fig. 2.28). That means that the same volume of structure is maintained throughout the internode, suggesting that the function and the activities within the internode remain constant, such as transferring the same volume of water.

The secondary cell wall of the xylem vessel consists of a thick wall containing lignin, which becomes thicker as the plant grows. Under a microscope, it was observed that the thick secondary wall of lignin surrounds the surface of the vessel tube, like a bellows, to provide mechanical support for its function of water or nutrient transport in the internode. Figure 2.29 shows an example of the vessels in a soybean plant. In the case of the thick wall, the bellows spacing became narrower with the development of the plant, suggesting that the volume of water leakage might

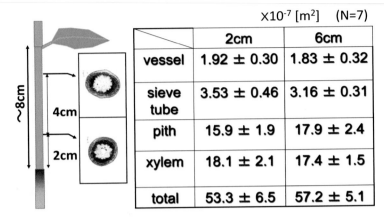

		2cm	6cm
	vessel	1.92 ± 0.30	1.83 ± 0.32
	sieve tube	3.53 ± 0.46	3.16 ± 0.31
	pith	15.9 ± 1.9	17.9 ± 2.4
	xylem	18.1 ± 2.1	17.4 ± 1.5
	total	53.3 ± 6.5	57.2 ± 5.1

×10⁻⁷ [m²]   (N=7)

Fig. 2.28 Tissue area in the section of the stem at different heights. The stem was harvested from 2 and 6 cm above the cotyledon of a soybean plant. The areas of the vessel, sieve tube, pith, and xylem were measured under a microscope. As shown in Fig. 2.27, although the shape and position of the xylem tissue varied, the area at different heights was approximately identical, which supports that the same amount of water was constantly transferring upwards

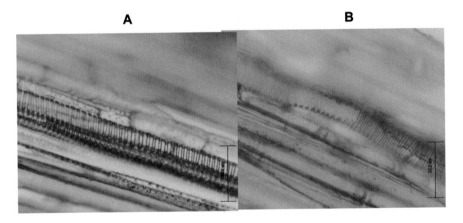

Fig. 2.29 Xylem structure under the microscope

be controlled by this spacing. Actually, at senescence, hardly any space for this bellows was observed surrounding the vessel (data not shown).

Since the flexible movement of the bordered pit field functions to adjust water movement, a question was raised about the water stream within the vessel: whether the water moves straight up or if there is any whirling motion in the vessel. To understand the upward flow of water within the internode, the Reynolds number in the vessel was calculated. It was a very small number, 0.002–0.003, given 20 μm as the diameter of the vessel, 1 mm/s as the water velocity and 10^{-6} m²/s as the kinematic viscosity, suggesting that the water flow within the vessel along the vessel wall was not creating any whirling stream.

Another question is whether there was isotope exchange between ^{15}O-water and natural ^{16}O-water in the vessel. To test this possibility, 10 ml of ^{15}O-water was passed through the column packed with 1.5 g of cellulose, the water coming out from the column was collected every 300 μL, and the ^{15}O-water concentration was measured. When isotope exchange occurs, the ^{15}O-water concentration in the collected fraction must be decreased. The ^{15}O-water decreased by 14% only in the first fraction, and there was hardly any decrease in the ^{15}O-water concentration in subsequent fractions. Therefore, isotope exchange within the internode was not taken into account in our experiment.

The motive force of water leakage from vessels and the subsequent lateral movement of water remains a major question. The pressure inside the vessel is low, approximately 0.8 M Pa. In spite of this low pressure, the water flowed out from the vessel horizontally. The pith might be one of the candidates for the horizontal route. As shown by the ^{3}H-water image after 20 s (Fig. 2.25), water moved horizontally toward the pith according to diffusion. Another question is the relation between the leaked water and that lost by transpiration, i.e., the relation between horizontal and vertical water flow.

Last, the water movement found here provided us with many further questions regarding how plants regulate the water flow within the plant, particularly what causes water leakage from xylem tissue and its return to the xylem, which seems to be an important and basic plant activity.

Bibliography

1. Nakanishi TM, Tanoi K, Yokota H, Kang DJ, Ishii R, Ishioka NS, Watanabe S, Osa A, Sekine T, Matsuhashi S, Ito T, Kume T, Uchida H, Tsuji A (2001) ^{18}F used as tracer to study water uptake and transport imaging of a cowpea plant. J Radioanal Nucl Chem 249(2):503–507
2. Nakanishi TM, Don-Jin K, Kitamura T, Ishii R, Matsubayashi M (1999) Identification of water storage tissue in the stem of cowpea plant (Vigna unguliculata Walp) by neutron radiography. J Radioanal Nucl Chem 242:353–359
3. Tanoi K, Hojo J, Nishioka M, Nakanishi TM, Suzuki K (2005) New technique to trace [O-15] water uptake in a living plant with an imaging plate and a BGO detector system. J Radioanal Nucl Chem 263:547–552
4. Ohya T, Tanoi K, Iikura H, Rai H, Nakanishi TM (2008) Effect of rhizosphere pH condition on cadmium movement in a soybean plant. J Radioanal Nucl Chem 275:247–251
5. Ohya T, Tanoi K, Hamada Y, Okabe H, Rai H, Hojo J, Suzuki K, Nakanishi TM (2008) An analysis of long-distance water transport in the soybean stem using (H$_2$O)-O-15. Plant Cell Physiol 49:718–729

Part 2
Elements in a Plant

When we consider the fundamental activity of plants, many questions remain: do we know the plant itself well? Plants require 17 elements. Do we know how the elements move or how the profile of the element change during the long-term of the developmental stage when the plant is growing under normal conditions, that is, how most plants live? When we regard the plant as an integrated system to let the elements work in the most efficient way, there must be some regulations even within the same tissue. Elements move according to the requirements of each tissue, since the age of each tissue within a plant is different, i.e. the plant consists of tissues at various stages, from meristem to senescent. In each tissue, the element concentrations or requirements are always changing. In mammalian tissue, for example, there are not such drastic changes in element concentration among different muscles as in plant tissues, because the growth stage of all the muscle tissue is about the same. The different concentrations of the elements in different tissues within the plant must reflect the plant-specific activities of life. An approach to determining the barriers or gradients of the element profiles or movement must be developed to understand the sophisticated regulation in a living plant.

An element absorbed by the root is transferred through the xylem and redistributed by the phloem with photosynthate, where the sieve tube is considered to regulate the element flow into tissues. However, the mechanism for the selective transport of chemicals in general is unclear. Additionally, the element flow varies with respect to the growth stage, especially at the flowering and fruit ripening stages. Since there is no well understood systematic approach to determine the kinetics of the element concentrations throughout the life cycle of a plant, to obtain fundamental knowledge of element profiles, neutron activation analysis was employed to measure the concentration of multiple elements in every tissue of the plant in detail throughout all developmental stages, from germination to seed ripening.

The advantage of neutron activation analysis to determine the elements in plant tissue is that this method provides the absolute amount of the elements. There is no other method to determine the absolute amounts of elements for two reasons. One is that neutron activation analysis can be performed nondestructively. Other sensitive analytical methods, such as inductively coupled plasma spectroscopy (ICP) and atomic absorption spectroscopy, require the sample to be prepared as a solution. That is, the sample must be dissolved in chemicals. However, even in highly purified reagents, trace amounts of other elements are present as contaminants, and this contamination from the reagent cannot be prevented. The second is that there is no way to measure how much of the sample is dissolved or what amounts of trace elements remain as a residue in solution, affecting the yield. In nondestructive analysis, there is no need to estimate how much of the sample remained undissolved.

Chapter 3
Element-Specific Distribution in a Plant

Keywords Element-specific distribution · Neutron activation analysis ·
Nondestructive element analysis · Barley · morning glory · ^{28}Mg · ^{28}Mg production ·
^{42}K · ^{42}K production · Short-day treatment · Dark/light cycle

3.1 Nondestructive Element Analysis: Element Profile

Neutron activation analysis (NAA) has high sensitivity, but the sensitivity is different among the elements. Table 3.1 shows the sensitivity of activation analysis to different elements. Among the elements, the sensitivity to the heavy elements is very high. We feel it is very difficult to convey how sensitive the analysis is. What should be considered in the actual experiment? One example that helps to illuminate the high sensitivity is that when people wearing gold accessories prepare irradiation samples, the gold vapor from the accessories contaminates the sample and is detected. Therefore, this high sensitivity for detecting elements has been applied for forensic studies.

A simplified description of the method of NAA is introducing the sample into a research reactor to produce radioactive nuclides from the corresponding stable nuclides contained in the sample. In the case of thermal neutron irradiation, elements in the sample produce radionuclides, mostly by (n, γ) reactions, resulting in one neutron-rich nuclide. Then, after irradiation, some of the newly produced radionuclides emit their own specific energy as γ-rays and can be detected by a γ-ray detector (Fig. 3.1). The kind of radionuclide produced is dependent on the nuclear property of the element and on the neutron energy and total flux of the neutrons irradiated. However, regardless of what clean sample is irradiated with thermal neutrons, not only plant samples but also mg levels of chemically washed small film or plastics, etc., ^{28}Al is the most likely radionuclide to be detected after irradiation, since the cross section, a kind of sensitivity, of the (n, γ) reaction of ^{27}Al is extremely high.

The measurement of radiation is as follows. After irradiation, the γ-rays emitted from the sample are measured by a γ-counter, such as a pure Ge counter or a Na(Tl)I scintillator, equipped with a pulse height analyzer. The kind of radionuclide produced was identified by the energy of γ-rays detected, and the amount of the element

© The Author(s) 2021   75
T. M. Nakanishi, *Novel Plant Imaging and Analysis*,
https://doi.org/10.1007/978-981-33-4992-6_3

Table 3.1 Sensitivity of the thermal neutron activation analysis

$(3{\sim}5) \times 10^{-13}$	Dy, Eu, In
$(1{\sim}4) \times 10^{-12}$	Co, Ag, Rh, V
$(6{\sim}9) \times 10^{-12}$	Mn, Br, I
$(2{\sim}5) \times 10^{-11}$	Th, Pr, Se, Lu, Nb, Ga, Sm, Cu, Re, Ho, U, Al, Hf
$(6{\sim}9) \times 10^{-11}$	Kr, Ba, Au, Ar, Cs
$(2{\sim}5) \times 10^{-10}$	Se, Er, Cl, W, Zn, As, La, Na, Pd, Pt, Yb, Gd, Ge
$(6{\sim}8) \times 10^{-10}$	Os, Te, Nd
$(1{\sim}3) \times 10^{-9}$	Tl, Rb, Sb, Sr, Ti, Mo, Xe, Mg, Cr, Hg, Y, Tm, K
$(4{\sim}7) \times 10^{-9}$	Ru, Sn, Tb, Ni, Ta, F, Ca
$(2{\sim}4) \times 10^{-8}$	Si, Ne, Ce, P, Cd
$(4{\sim}5) \times 10^{-7}$	S, Bi
$(2{\sim}5) \times 10^{-6}$	Zr, Pb, Fe
$(2{\sim}6) \times 10^{-4}$	O, N
0.11, 0.33, 19.2	Be, H, C

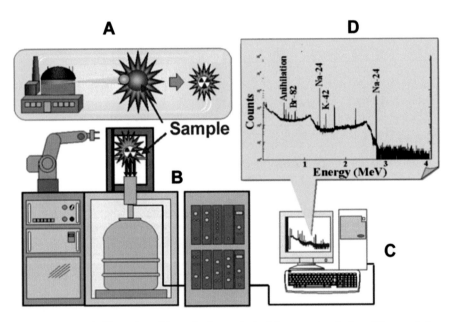

Fig. 3.1 Schematic illustration of the activation analysis. (**a**) neutron irradiation of the sample in a research reactor; (**b**) γ-ray counting of the irradiated sample by a Ge detector; (**c**) γ-ray spectroscopy by a computer; (**d**) example of a γ-ray spectrum

was calculated from the intensity of the γ-rays. However, the half-lives of newly produced radionuclides are different; since some of the radionuclides produced have very short half-lives, they rapidly decay out during the measurement. Therefore, γ-ray measurement should be performed at an appropriate time after irradiation checking the changes in the γ-ray spectrum.

For example, when a small amount of plant tissue is irradiated, a high amount of ^{28}Al is produced from ^{27}Al by the ^{27}Al $(n, \gamma)^{28}$Al reaction, and the half-life of ^{28}Al is only 2 min (γ-ray energy: 1.779 MeV). Though ^{28}Al decays out rapidly during the measurement, while a high intensity of radiation from ^{28}Al emits, the background level of γ-ray measurement is high, and this high background prevents the measurement of the γ-rays emitted by other radionuclides, especially those whose γ-ray energies are lower than approximately 1.5 MeV, in this case.

With the decay of ^{28}Al, the γ-rays of other nuclides with longer half-lives than that of ^{28}Al, such as ^{24}Na (1.369 MeV), ^{27}Mg (1.014 MeV), and ^{42}K (1.525 MeV), emerge in the γ-ray spectrum during measurement, revealing γ-ray peaks previously hidden under the overwhelming intensity of the ^{28}A γ-rays. From the counts of each nuclide-specific γ-ray, the absolute amount of the original element can be calculated by comparison to the radiation from a standard sample irradiated at the same time.

3.1.1 Profile of the Elements in Barley

The profile of the elements in barley leaves was obtained by employing NAA as follows. Barley (*Hordeum vulgare* c.v. Minorimugi) seed was germinated, grown in water culture, and harvested after 12, 19, 23, and 46 days. Then, 2 cm samples of the tip, middle, and bottom sections of each leaf were cut out, and neutron activation was performed. Each sample was sealed doubly in a well-washed polyethylene vinyl bag and then irradiated by a Triga Mark II type atomic reactor at the Institute for Atomic Energy, Rikkyo University in Japan, now decommissioned, with a thermal neutron flux of 1.3×10^{12} n/cm^2/s for 3 min. After the sample cooled for 3 min, the γ-rays emitted from each sample were measured by a Ge(Li) detector for 200 s. To determine the P concentration, the samples were placed in a Cd container for irradiation to measure only ^{28}Al produced by the fast neutron reaction of (n, α) from ^{31}P. Since ^{28}Al is also produced by the thermal neutron reaction of (n, γ) from natural ^{27}Al, a Cd container was needed to prevent this thermal neutron reaction. From this short-time activation analysis, ^{24}Na, ^{27}Mg, ^{28}Al, ^{38}Cl, ^{42}K, ^{40}Ca, and ^{56}Mn were produced, and their radioactivity counts corresponded to the amounts of Na, Mg, P, Cl, K, Ca, and Mn, respectively. To determine the element amount, reference materials, namely, JB3 (Geological Survey of Japan) and orchard leaves (National Bureau of Standards, present NIST, USA), were irradiated at the same time.

Figure 3.2 shows the distributions of these elements within a leaf during the developmental stage. The leaf number was added successively from the first leaf until the seventh leaf, which emerged on day 46. Two types of element distribution patterns for the highest accumulation in a leaf. One is that the content of the element decreased with the leaf number (Na, Mg, Cl, and Ca), and the other is that a maximum element concentration (P, K, and Mn) occurred. The element profile was determined in the same way throughout the developmental stage. The results of all the elements showed that in the 46-day-old sample, the element concentrations were drastically increased from those in the 31-day-old sample.

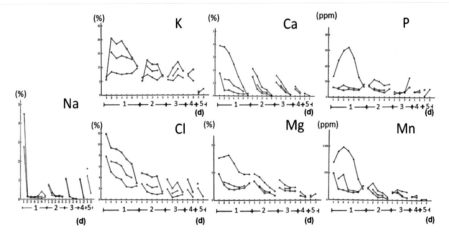

Fig. 3.2 Changes in Na, Mg, P, Cl, K, Ca, and Mn concentrations in barley leaves [1]. Large numbers 1–5 refer to the growth stage of the sample; 1, 2, 3, 4, and 5 correspond to 46 days, 31 days, 23 days, 19 days, and 12 days after germination, respectively. Small numbers 1–7 refer to the leaf number at each growth stage. Leaf number 1 is the eldest leaf. There were 2 leaves at 12 (5) and 19 (4) days and 7 (1) leaves after 46 days. △, ○, and □ denote the element concentrations at the tip, middle, and bottom parts of the leaf, respectively. The vertical axis indicates each element concentration in tissue (5). The concentration of the elements in each tissue was measured by the neutron activation analysis

There was an extremely high accumulation of Na in the first leaf, especially at the bottom part, whose concentration on the 46th day was 20 times higher than those of the other leaves. However, in other leaves, the Na concentration remained rather uniform and low, regardless of the leaf number and the position in the leaf. When the profiles of the 46-day sample of Na were compared to those of K, the tendency was that the Na content in the bottom part of the leaf was the highest in the first leaf and slightly higher in younger leaves, showing the reverse distribution from that of K. Considering this reciprocal tendency of Na concentration to that of K, as well as the high concentration of Na when K was eliminated from the culture medium (data not shown), there might be some compensatory function between K and Na content in the leaf. However, despite the same cation valency number, the distribution patterns of K and Na were drastically different.

The elements Cl and K are the only ones that are not constituents of organic structures but function mainly in osmoregulation. The concentration change within one leaf took place gradually, without any drastic accumulation at one edge of the leaf; therefore, the K and Cl contents in the middle part of the leaf were the average of the amounts in the tip and bottom parts of the leaves. However, it is interesting that Cl, K, and Na, which are known to regulate the osmotic pressure of cells, tended to accumulate in the bottom part of the leaf, which might suggest fast movement to the other tissues.

It is also interesting that K, which is the most abundant cation in the cytoplasm and whose cytoplasmic concentration remains in a relatively narrow range to stabilize the pH concentration, had a large concentration gradient in the leaf.

When the element moves along with the transpiration stream and redistributes to the other tissues smoothly, the element abundance in the middle of the leaf could be an average value of those in the tip and the bottom part. Extremely different concentrations of the element within a leaf might suggest some barriers to movement or fixation of the element at certain sites. A steep gradient in the element concentration within a leaf was shown for P and Mn, where the concentration was especially high at the tip of the leaf. P serves as a constituent of proteins and nucleic acids; therefore, it might not be released readily from the tip. Sometimes phosphate movement is considered to be similar to water movement; therefore, a high concentration of P at the leaf tip might reflect the result of intensive water evaporation. However, although P is known to undergo fast chemical changes in plants, the exact chemical form in which it is fixed is uncertain from this analysis. Regarding the chemical form of P in a plant, although it should be discussed later, we found that ^{32}P was in the form of phosphate, at least for the first 30 min, after being absorbed from the roots, in *Lotus japonicus*.

Calcium is the only element known to function mainly outside the cytoplasm, and its uptake rate into the cytoplasm is strictly restricted. A concentration of Ca a few times higher at the leaf tip might be explained by the low mobility of Ca from cell to cell. The concentration profile of Ca was similar to that of Mg. Although the behaviors of these elements are expected to be similar because they are in the same group in the periodic table, Mg has its own specific characteristics. One of them is that the Mg concentration is known to be kept constant among different plant species, suggesting a role in maintaining homeostasis in plants.

The extremely low concentrations in the senescent leaves, namely, the 46-day-old samples, suggested that the elements were recycled or transferred within the plant and moved toward another tissue where these elements were needed. Actually, in the case of a rice plant, it is known that N is transferred from senescent leaves to other tissues and used again, which might be regarded as an element recycling activity.

3.1.2 Profile of the Elements in Morning Glory During Growth

Another example of the element distribution is the Japanese morning glory (*Pharbitis nil* c.v. Violet). In this case, each leaf, internode, and root were harvested during the developmental stage, and neutron activation analysis (NAA) was performed. The method of preparation of the sample and irradiation is the same as that described for barley. Figures 3.3, 3.4, 3.5 and 3.6 show a schematic illustration of the element profile in morning glory grown in soil at each stage from germination to seed ripening. The element concentrations of tissues were classified into more

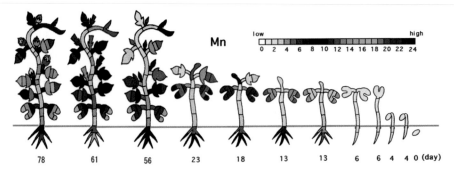

Fig. 3.3 Schematic illustration of the Mn concentration profile in a morning glory during the developmental stage [2]. The plants were harvested from 0 to 78 days after germination, and all tissues in a sample were separated. The concentrations of the elements in each tissue were measured by the neutron activation analysis and classified into 24 grades, and pseudocolors were assigned

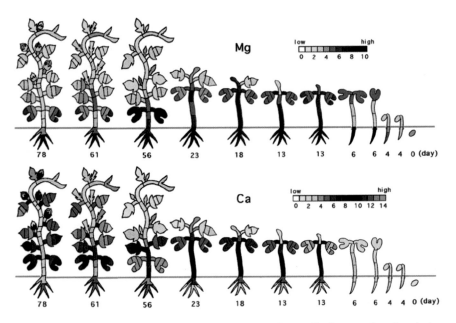

Fig. 3.4 Schematic illustrations of the Mg and Ca concentration profiles in a morning glory during the developmental stage [2]. The concentration of the elements in each tissue was measured by the neutron activation analysis, and pseudocolors were assigned

than 10 grades and were assigned to the corresponding level of the pseudocolor scale.

The morning glory bears the flower bulb, flower, and seeds after 56, 61, and 78 days of germination, respectively. The plant developmental stage from the juvenile phase to the adult phase occurs between 23 days and 56 days. As shown

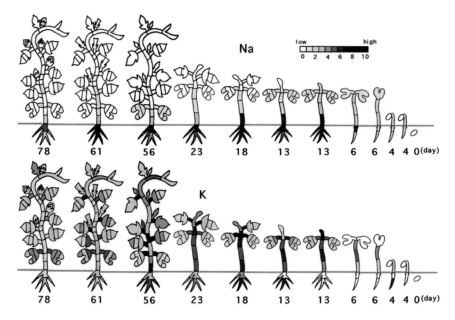

Fig. 3.5 Schematic illustrations of the Na and K concentration profiles in a morning glory during the developmental stage [2]. The concentration of the elements in each tissue was measured by the neutron activation analysis, and pseudocolors were assigned

in the figures, systematic barriers appeared according to each element profile. The first barrier was found between the root and the upper part of the plant.

Among the elements, the profile of Mn was representative, showing a high concentration in senescent tissue, especially in the elderly leaves, suggesting rapid movement of the element to the senescent tissue with growth. Mn seemed to move with the water in xylem tissue and then accumulate at the leaf tip (Fig. 3.3).

The concentration profiles of Mg and Ca (Fig. 3.4) showed similar distribution patterns, where the highest concentration was found below the cotyledon before flowering was induced. With the development of the plant, there appeared to be a distinct difference in element distribution between the juvenile phase and adult phase. When the flower bulb was developed, on day 56, the stored Mg and Ca in the lower tissue were moved to the upper part of the cotyledon.

Although both alkaline elements Na and K produce monovalent cations, the profiles were drastically different (Fig. 3.5). Most Na accumulated in roots, and the concentration of Na in the aboveground part was very low, whereas most of the K absorbed was transferred to the aboveground part. The reciprocal distribution of Na and K was also observed in barley (Fig. 3.2).

A barrier around the cotyledon was also shown for Cl and Br (Fig. 3.6) but did not disappear completely until the seed ripening stage. The halogen elements Cl and Br are volatile and easily lost from the plants during the developmental stage; however,

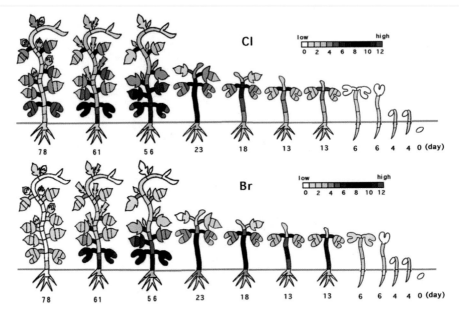

Fig. 3.6 Schematic illustrations of the Cl and Br concentration profiles in a morning glory during the developmental stage [2]. The concentration of the elements in each tissue was measured by the neutron activation analysis, and pseudocolors were assigned

higher concentrations of Cl and Br were still found in the petiole of the cotyledon and the first leaf at the senescent stage.

When the elemental concentration in the internode was plotted, the difference in concentration between tissues, indicating a barrier, was shown more clearly. Figure 3.7 shows the concentration ratio of the element in the leaf petiole between the leaf stem (LS) and the connecting leaf (L). When the ratio is greater than 1, the concentration of the element in the petiole is higher than that in the connecting leaf, suggesting that the high concentration of the element reflects a barrier to the movement of the element toward the leaf at the connecting site.

In the case of K, the concentration at the leaf petiole was always high during development, indicating that the high concentration of K at the leaf petiole always exerted pressure on the element to move toward the connecting leaf. This ratio tended to increase in the younger leaves, suggesting the regulation of the movement of the elements to accumulate in younger leaves. In addition to the leaf, the concentration of K in the petiole was also higher than that in the connecting internode, especially during flower formation. Therefore, it was suggested that the regulation of K movement in both directions, toward the connecting leaf and the internode, indicated some role in the petiole (Fig. 3.5). This tendency toward higher concentrations at the petiole than at the connecting leaf and internode was also observed for the elements Ca, Cl, and Br (Figs. 3.4 and 3.6).

The concentration of the elements at each internode within the same stem during the developmental stage is plotted in Fig. 3.8. The concentrations of the elements K,

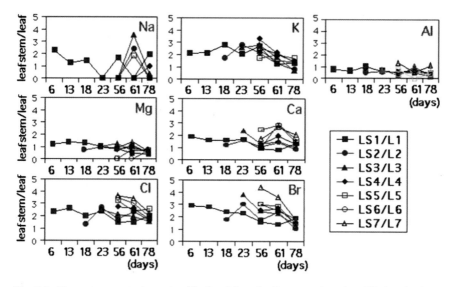

Fig. 3.7 Element concentration ratio of leaf petiole to leaf in a morning glory [2]. A ratio above 1 shows that the element concentration in the leaf petiole is higher than that in the leaf. LS: leaf stem; L: leaf

Mg, Ca, Cl, and Br increased gradually from the juvenile phases and showed a maximum during the adult phase. Then, the concentration decreased toward the senescent phase. However, in the case of Al and Na, the maximum concentration appeared at a very early stage, at approximately 6 and 13 days, and most of these elements accumulated in roots and did not move to the aboveground part of the plant.

In the case of the seed maturing process, there was a clear element partition among the seed tissues, calyx, seed coat, seed wall, or endosperm. In the mature seed, where the seed wall was well developed, it was interesting that the element selectively accumulated in the seed wall, and only a small amount was partitioned into the embryo or endosperm (data not shown).

What about the heavy element profiles in the plant? Figure 3.9 shows the concentration profiles of Al and V during the developmental stage. They stayed in the roots and did not move to the aboveground part throughout the growth. Generally, most of the heavy elements tend to accumulate in roots and were hardly transferred to the aboveground part, except for Cr and Mn, according to our observations. Figure 3.10 shows the heavy element distribution throughout the plant after 78 days of germination-bearing seeds. As shown in the figure, it is not known why two elements, Cr and Mn, moved to the aboveground part. One of the reasons for Mn might be that it is required to react with chlorophyll in photosynthesis to produce O_2. Another explanation for the movement might be that there are many valences for these ions, from +2 to +7 for Mn and + 2, +3, and + 6 for Cr, and these variable valences might facilitate chemical bonding or reactions leading to movement.

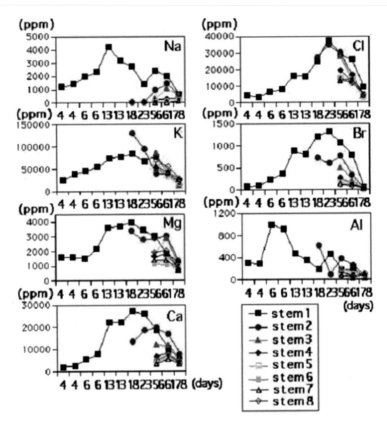

Fig. 3.8 Profile of element concentration in each stem of a morning glory [2]. The number of stems is counted between two nodes from the lower stem to the upper stem toward the shoot, where stem 1 is the stem between the root and cotyledon

To determine the profiles of the heavy elements in more detail, root concentration was omitted, and the concentrations of the elements in the aboveground part alone were graded into 15 steps (Fig. 3.11). There appeared to be a specific feature of the Co profile where the concentration in the internode was higher than that in the other tissues. This kind of profile was not observed in other elements. It was known that the concentration of Co in the fertile pasture for growing live stocks was high, and this feature might, therefore, be derived from the Co profile in the grass.

A summary of the distribution of the elements during growth is as follows.

1. Juvenile phase and adult phase

   During the juvenile phase, Ca and Mg accumulated below the cotyledon, and then, when the plant reached the adult phase, they moved upward. In the internode, these element concentrations increased during the juvenile phase and decreased during the adult phase.

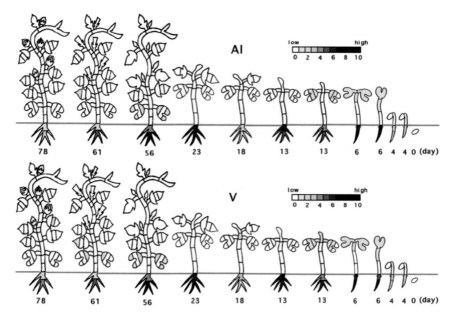

Fig. 3.9 Schematic illustration of Al and V concentration profiles in a morning glory during the developmental stage. The concentration of the elements in each tissue was measured by the neutron activation analysis, and pseudocolors were assigned

2. Element barriers in the tissue

The barrier between roots and the aboveground part of the plant was high, especially for Na, Al, and heavy elements other than Cr and Mn. The concentrations of K, Ca, Mg, Cl and Cl in the leaf petiole were higher than those in the connecting internodes and leaves.

3. Seed maturing process

During the maturing process, the element concentration ratio of seed to seed stem increased. When the seed was mature, most of the elements occurred predominantly in the seed coat or seed wall, not in the embryo or endosperm.

3.1.3 Profile of the Elements in Young Seedlings of Morning Glory

The overall element profile during 78 days of development is shown. The next analysis presented is the change in the elemental profile within a few days, with a connection to circadian rhythm. The circadian rhythm plays an important role in plant development, such as stem elongation and leaf movement according to light conditions. A Japanese morning glory (*Pharbitis nil.* cv. violet) is known to be very sensitive to light conditions, and a single treatment of seedlings with an increased dark period, from 8 to 16 h, can induce flowering. Although it is known that the

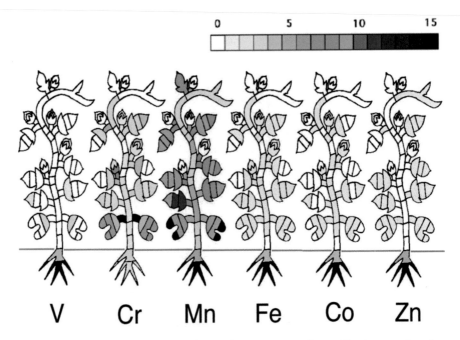

Fig. 3.10 Schematic illustration of the heavy element concentration profiles in a morning glory after 78 days of germination [3]. Pseudocolors were assigned. Most heavy elements accumulated in the roots except Cr and Mn

circadian rhythm of light acts to promote floral induction in the shoot apical meristem, through the FT signaling system (expression of flowering locus T (FT) gene), the elemental profile at flower induction in the meristem is not well known.

Using the same kind of morning glory, the circadian rhythm of elemental concentration was studied. First, the concentration profile in the seedling was analyzed after growth for approximately a week after germination in water culture under 12 h L/12 h D light/dark conditions. The seedlings were periodically harvested and separated into 9 tissues: cotyledon, petiole, shoot apex, and three parts each of the stem and roots, upper, middle, and bottom. To determine the amount of the elements, neutron activation analysis with γ-ray spectroscopy was performed using the research reactor JRR3M installed at the Japan Atomic Energy Agency (JAEA). Each sample was sealed in an ultrapure polyethylene bag and irradiated for 10 s. The total thermal neutron dose was 1.9×10^{14} n/cm^2. After irradiation, the sample was cooled for 2 min, and the gamma-rays from the sample were measured for 150 s by a pure Ge counter.

Figure 3.12 shows the concentration profile of 5 elements in the 7-day-old seedlings. A definite distribution pattern of the element was observed even in the 7-day-old seedlings, and the tendency was the same as those shown in Fig. 3.12. The largest amounts of Na and K accumulated in the roots, and the other elements, Mg,

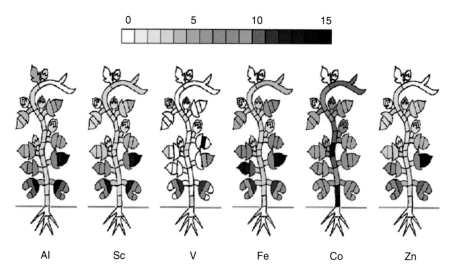

Fig. 3.11 Schematic illustration of the elemental profiles in the aboveground part of a morning glory after 78 days of germination [3]. The elemental concentrations of the aboveground parts of the plant tissues, i.e., all parts except the roots, were classified into 15 grades and assigned to the corresponding pseudocolor. The maximum concentrations (ppm) of Al, Sc, V, Fe, Co, and Zn were 723, 0.102, 1.20, 119, 1.32, and 0.426, respectively, whereas the maximum concentrations (ppm) in the roots were 4020, 2.03, 6.44, 1240, 23.0, and 7.39, respectively. Since the concentration data of the roots were omitted, no root was colored

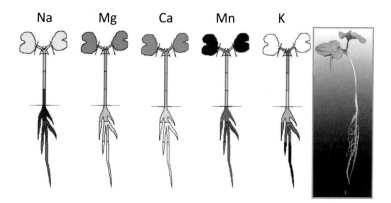

Fig. 3.12 Schematic illustration of the elemental profiles in morning glory after 7 days of germination. The concentration of the elements in each tissue was measured by the neutron activation analysis, and pseudocolors were assigned

Ca, and Mn, spread upward to the other tissues. The K concentration was high at the root tip and gradually decreased toward the upper part, and K accumulated in the internode but not in the leaves. The amount of potassium in the plant was very high, approximately 0.3 to 8% of the total fresh weight of the plant. In the case of Mg and Mn, their concentrations were high the in leaves because of photosynthesis

requirement. Ca and Mg always had a similar concentration profile between Ca and
Mg, but the Ca concentration was always approximately 1.5 times higher than that of
Mg.

3.1.4 Ca and Mg Concentrations

Each element showed its specific distribution pattern within a seeding, as shown
above. Although the macroscopic concentration pattern of the element did not
change during growth, on a finer level, the concentration within each tissue was
found to change with hours. To study the change in elemental concentration with
respect to the light conditions, the seedlings were harvested periodically during the
water culture, and the elements in each tissue were measured in the same way by
neutron activation analysis (NAA).

Among 5 elements investigated in the seedlings, Ca and Mg showed changes in
their concentrations with respect to the light conditions, especially at the shoot apex.
The concentrations of both elements at the shoot apex tended to increase during the
light period and decrease or remain constant during the dark period. As shown in
Figs. 3.13 and 3.14, the concentrations of both elements were distinctively different
between the shoot and roots.

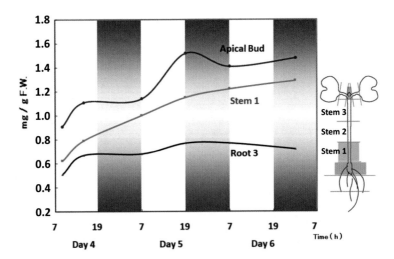

Fig. 3.13 Concentration of Ca under normal conditions [4]. Vertical color columns from 19 h to
7 h are the dark periods. Stem 1: bottom part of the stem; Root 3: upper part of the root adjacent to
Stem 1, as illustrated. There was a circadian change in Ca concentration, where the concentrations
in the apical bud and root 3 increased during the light period and decreased or did not change during
the dark period. The Ca concentration in Stem I continuously increased in different light conditions;
however, the concentration in Stem 1 was twice that in the neighboring tissue, i.e. Root 3

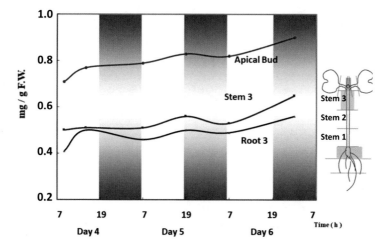

Fig. 3.14 Concentration of Mg under normal conditions [4]. Vertical color columns from 19 h to 7 h are the dark periods. Stem 3: upper part of the stem, adjacent to the apical bud; Root 3: upper part of the root, connecting to the bottom part of the stem, as illustrated. The circadian change in Mg concentration was observed in all three tissues: the concentration increased during the light period and decreased during the dark period. There is a large gap in concentration between neighboring tissues (Apical Bud and Stem 3), where the Mg concentration in Stem 3 is approximately 60% of that in the Apical Bud

Since the Ca concentrations in the shoot apex, stem 3 connected to the shoot apex, and stem 2 and stem 1 showed similar levels (data omitted), approximately 2 times higher than that in root 3, it was suggested that the Ca concentration was maintained at the same level among the organs above the ground.

Compared to the Ca distribution, there was a distinctive gap in Mg concentration between the shoot apex and stem 3, although these tissues were connected to each other. The magnesium concentration in stem 3 was only approximately 60% of that in the shoot apex. Considering the different positions of the concentration gaps of Ca and Mg, between roots and stems, and between stems and the shoot apex, respectively, it was suggested that the Mg concentration was more severely regulated than that of Ca.

A clear difference in the concentrations of the two elements was found when short-day treatment was performed from day 5 (Figs. 3.15 and 3.16). When the growth condition was changed to the short-day period, from 15 h on day 5, the concentration rhythm depending on the light–dark condition was lost. Although the Mg concentration increased gradually in all organs during the light period, after short-day conditions were introduced, the Mg concentration continued to increase even in the dark period until the second dark period of the short-day treatment and then suddenly decreased to almost the same concentration level as that at 7 h of day 5. In stem 3 and root 3, the tendency was similar to that of the shoot apex, except for the earlier appearance of the turning point to begin the decrease.

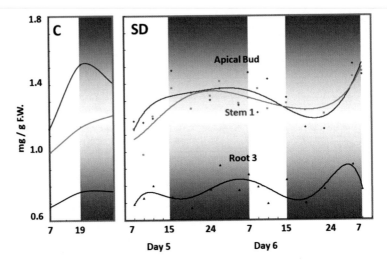

Fig. 3.15 Concentration of Ca under the short-day treatment [4]. Five days after germination, a dark period was introduced from 15 h. Left: Ca concentration on day 5 under normal conditions. C: control; SD: short-day treatment

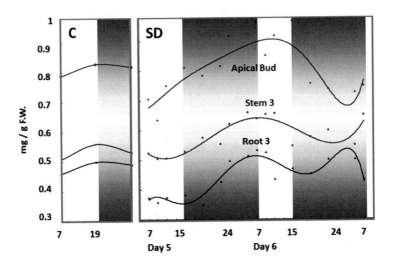

Fig. 3.16 Concentration of Mg under the short-day treatment [4]. Five days after germination, a dark period was introduced from 15 h. Left: Mg concentration on day 5 under normal conditions. C: control; SD: short-day treatment

In the case of Ca, the increase in Ca concentration stopped at the beginning of the short-day treatment, plateaued during the first dark period, and then decreased during the light period of day 6 in the shoot apex as well as stem 1, suggesting that the Ca concentration was maintained at the same level in both tissues. After day 7, the diurnal rhythm in the concentration appeared again. The large concentration gap between stem 1 and root 3 indicates that there is some mechanism that controls the Ca distribution between the bottom of the stem and the root.

Then, what could be suggested by the different changes in the concentrations of Mg and Ca? The changes in Ca and Mg concentrations showed similar tendencies during ordinary light–dark conditions throughout the whole plant, especially in the shoot apex. However, a clear difference between the two elements was found under short-day treatment. After the first longer period of darkness, 16 h, during the 8 h light period on day 6, the concentration of Ca in the shoot apex decreased, while that of Mg increased, which might indicate a compensatory role of Mg for decreased Ca, for example, to maintain the proper pH value and avoid the accumulation of organic acids. The increase in concentration during the light period could be well explained by transpiration, the major driving force of mass flow in the xylem. The increase during the dark period, on the other hand, could be understood only when endogenous "circadian rhythms" were taken into consideration. That is, diurnal concentration change was lost during the response to the short-day treatment and then "reset" again to the circadian rhythms even during the dark period.

Although Mg has not been mentioned to have any effect on flower induction before, our data suggested that Mg had some role in the shoot apex where the new leaves and bulbs appear. Therefore, after showing the circadian rhythms of Mg and Ca by NAA, the following fluorescent study was performed. The part of this study related to fluorescence imaging is introduced briefly below. It was difficult to distinguish the differences between these elements by the fluorescent staining method, not only because of the similar chemical behavior between the elements but also because the image of Mg^{2+} was always hidden by the Ca^{2+} image due to the overwhelming abundance of Ca^{2+}. We were able to show the different distributions of Ca^{2+} and Mg^{2+} by employing two fluorescent probes, Mag-fluo-4 AM and Fluo-3 AM (Molecular Probes, Inc., Eugene, Oregon). The Mag-fluo-4 AM probe was originally designed for binding Ca^{2+} but modified to be additionally responsive to Mg^{2+}. Using two probes, the distribution of these elements at the apical meristem, where the flower buds emerge, was visualized (Fig. 3.17).

During growth in the vegetative phase, cells in the center of the top layers accumulated high amounts of Mg^{2+}. Exposure to a single short-day treatment induced the flowering process and dramatically reduced the fluorescence associated with Mg^{2+} accumulation in the top layers, suggesting that Mg^{2+} contributed to the flower induction process. The fluorescence associated with Ca^{2+} did not show this distribution difference before and after the short-day treatment. A night break treatment also showed a similar Mg fluorescence pattern. Since Mg^{2+} might play an important role in flower induction (Fig. 3.17), this Mg study was further developed to produce ^{28}Mg tracers and to develop a real-time RI imaging system (see next Sect. 3.2).

The flowering process is crucial for plant development. Plants begin life in the vegetative phase and shift to a reproductive phase depending on environmental signals or aging. In the case of morning glory, one hundred percent of flowering was attained when removal of the cotyledon took place 2 h after the end of the short-

A   **B**

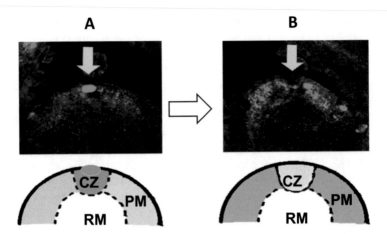

Fig. 3.17 Mg distribution in the shoot apical meristem of morning glory [5]. Confocal laser scanning microscope images of shoot apical meristem stained with Mag-fluo-4 AM before (A) and after (B) the short-day treatment with schematic illustration. High Mg2+ concentration was localized at the central zone (CZ) in meristems before a flower induction treatment with a long dark period. After flower induction, this high concentration disappeared. This drastic change in distribution of Mg^{2+} was not observed in other tissues. PM: peripheral meristem; RM: rib meristem

day treatment, indicating that the flowering signal(s) reached the shoot apical meristem at this time (data not shown).

Although the flowering process is not well understood, flowering is reported to be controlled by endogenous diurnal rhythms mediated by phytochrome and pH, which show oscillating patterns, as well as abscisic acid (ABA), which affects photoperiodic flowering in relation to a dark period. The time course and the movement of some components are key to understanding the mechanism of flowering. Our fluorescent staining study showed that there was a specific accumulation of Mg in the shoot apex, and this accumulation disappeared during flowering induction.

3.1.5 Al Concentration

Finally, the change in the Al concentration in the root tip is presented. Aluminum is a toxic element to plants, and most studies focus on the negative effect of Al on root tips when the Al content in soil increases. However, the Al concentration naturally contained in roots and the changes in Al content have attracted less attention. Since the detection limit of Al in neutron activation analysis is extremely high, the Al concentration in the root of morning glory was analyzed under normal conditions without the addition of any Al chemicals in the culture solution.

Figure 3.18 shows the change in the concentration of Al in the root tip of morning glory from the fourth and fifth days after germination. Since the Al concentration in other parts of the root was lower than 0.004 mg/g F.W., only the Al concentration in

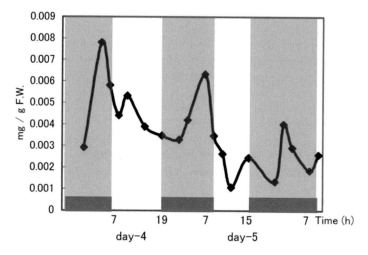

Fig. 3.18 Concentration of aluminum in the root tip [4]. The condition was changed to short-day from 15 h after 5 days of germination. The thick gray bars on the horizontal axis show the dark periods

the root tip was plotted. Aluminum ions were not added to the nutrient solution; therefore, the detected Al seemed to originate from the seed itself, and the Al concentration in seeds was the same as that in the root tip (data not shown). The overall Al concentration in the root tip was high at the younger stage and gradually decreased with development; however, the concentration changed periodically under normal light conditions. That is, with time, the concentration in the root tip became much higher than that in the other parts of the root. However, these Al concentration peaks appeared during the dark period, unlike those of Ca and Mg. Furthermore, peaks appeared approximately 10 h after the beginning of the dark period, which could be interpreted as a few hours before the beginning of the light period, dawn, as if the seedling knew when to expect the sunrise. After the short-day treatment, the concentration peak still appeared at a similar timing to that before the short-day treatment, and then this concentration timing changed further. The concentration pattern suggested that there was always a particular time for the secretion of trace amounts of Al from roots. It is not known whether the high peak of Al is attributable to absorption or to the relocation of Al in the root.

3.1.6   Summary of NAA

By applying neutron activation analysis, the absolute amounts of the elements were determined nondestructively in both barley and morning glory. Each element has its own distribution pattern, suggesting a specific physiological role of each element in different tissues of the plant. It seemed that there are many concentration junctions for each element to pass through, and these junctions exist at every connection between tissues in the whole plant. This distribution pattern was maintained throughout the developmental stages. The element-specific distribution pattern of barley was similar to that of morning glory, suggesting that the roles of each element in different plants were similar. Heavy elements, except for Cr and Mn, accumulated in roots and hardly moved to the aboveground part of the plant. Although the macroscopic pattern of the element distribution in tissues showed a similar tendency during growth, the element profiles changed in the course of hours, especially under different light conditions. Among the elements studied, Ca and Mg showed circadian rhythms in their shoot concentrations, increasing during the day and decreasing during the night. There was also a rhythmic change in the Al concentration in the root tip according to the dark/light conditions.

Since plant nutrients are mainly inorganic ions, elemental movement is expected to provide some clues to analyze the physiological development of plants. These findings triggered the further development of a real-time element-specific imaging system using RI (See Part II, Chap. 4).

3.2   Radioactive Nuclide Production for Mg and K

Although plants require 17 elements, the plant physiology of the elements for which radioisotopes are not available, such as B or Si, an essential element and a useful element, respectively, has not been well studied.

Figure 3.19 shows 17 essential elements and 5 useful elements. As is known, without essential elements, plants cannot grow normally or cannot complete the developmental stage, and without useful elements, the appropriate yield cannot be

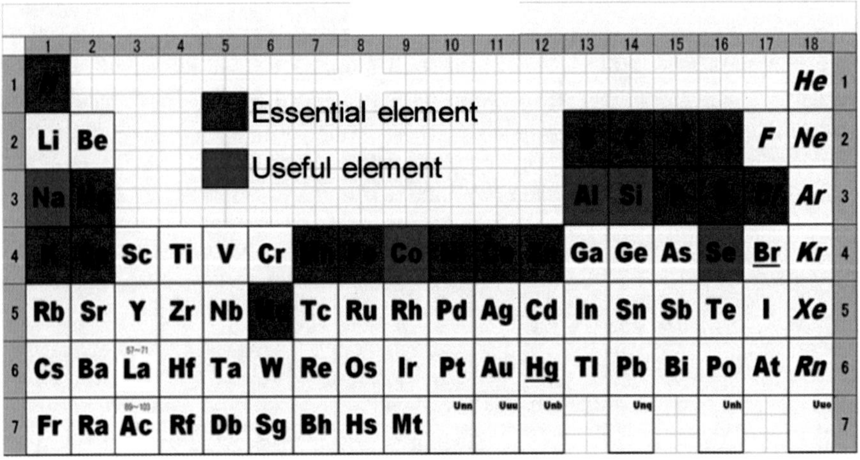

Fig. 3.19 Essential and useful elements for plant growth

Fig. 3.20 Element
concentration in a plant and
fertilizer. (**a**) average
concentration of mineral
nutrients in plant shoot dry
matter; (**b**) element
concentration of a typical
chemical fertilizer
(14:14:14))

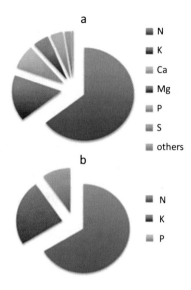

expected. These properties are essentially based on the growth of agricultural
products. From this point of view, the ratio of the essential elements contained in a
plant and the ratio of the elements prepared in a representative chemical fertilizer are
similar (Fig. 3.20). Since plants require inorganic elements to grow, the physiological properties of each element are among their most important factors and have been
studied utilizing radioactive nuclides as tracers.

Because of the lack of radioactive nuclides available for tracer work, the physiological study of some elements, such as B or Si, is far behind that of other elements.
For example, the physiological study of B has begun to develop only after the recent
identification of the transporter of B, BOR1 in Arabidopsis, owing to the development of an analytical method for determining trace amounts of B. Another difficulty
in studying B was the difficulty out of determining the chemical reactions of B in
plants. Although B reacts with the chemical configuration of cis –OH groups, such as
marine plants, the sugars of higher plants do not have cis –OH groups. Therefore, the
role of B in higher plants was difficult to pursue. In the case of Si, a highly sensitive
analytical method has not yet been developed. However, the abundance of Si in the
surface of the Earth is 25.8%, which is extremely high, second only to O, which has
an elemental abundance of 49.5%, according to the Clarke number. Although the
amounts of both elements, B and Si, can now be measured by inductively coupled
plasma-mass spectrometry (ICP-MS) or optical emission spectrometry (ICP-OES), it
is still difficult to analyze and trace the behavior of these elements at the level of
radioactive tracer work.

There are still three other essential elements, O, Mg, and K, for which radioactive
nuclides are not commercially available, because of the short half-lives of the
corresponding radioactive nuclides, ^{15}O, ^{28}Mg, and ^{38}K or ^{42}K. In the case of ^{15}O,
we produced and employed ^{15}O-labeled water to analyze water movement as

described in Part I, Chap. 2, Sect. 2.3. Because of the extremely short half-life and rapid decay of ^{15}O, 2 min, the nuclide had to be prepared immediately prior to the experiment, and a large amount of ^{15}O was employed to conduct the experiment. The difficulty of preparing ^{15}O-labeled compounds within a short time before starting the experiment is another reason the physiology of ^{15}O-labeled compounds in a plant has been little studied. In addition to ^{15}O, we prepared ^{28}Mg and ^{42}K ourselves and were able to perform radioactive tracer work before the nuclides decayed out. In the following sections, the preparation of Mg, with a short introduction to tracer work using ^{28}Mg, and the production of ^{42}K tracers are presented.

3.2.1 Production of ^{28}Mg

First, we briefly introduce the features of Mg before describing the preparation of ^{28}Mg. Although Mg is an essential element for plant growth, since Mg is a component of chlorophyll to carry out photosynthesis, the radioactive tracer was not applied to study plant physiology. Magnesium is the second most abundant cation in living cells, and over 300 enzymes are known to be Mg dependent. Mg concentration changes significantly affect the membrane potential. The important feature of Mg is that the concentration of Mg in plants is constant, within a small range, among different plant species; therefore, Mg is estimated to have a role in maintaining the homeostasis of plant physiological activity.

When Mg is deficient in soil, such as at low pH and high concentrations of K^+ or NH_4^+, chlorosis occurs, especially in young mature leaves, changing the green color to yellow. Mg deficiency is also liable to appear during the ripening process of the fruit and decrease the yield. The mechanism is estimated to be closely related to sugar transport. The accumulation of Mg in the fruit drives Mg deficiency in leaves, which induces the deactivation of sugar transport in the phloem and inhibits translocation.

There are two strategies by which plants address element deficiency. One is the retranslocation of the element to maintain growth, which requires mostly phloem transport of the element from mature tissues. The other is to increase the uptake of deficient elements by the roots, such as by inducing the expression of root transporters of the element or secreting chemicals to produce compounds of the element.

In the case of Mg, the early response to Mg deficiency is now an important issue, i.e. what occurs before the accumulation of sugars and subsequent reactions to reduce photosynthesis or to increase the heavy element concentration, which induces reactive oxygen species and chlorosis, etc. Because of the lack of available Mg tracers, Co or Ni is sometimes used to replace Mg, but the physical properties of Mg cannot be not well determined by using these elements.

To perform Mg tracer work, the useful radioactive nuclides for Mg should be taken into account. There are two candidates for Mg tracers, ^{27}Mg and ^{28}Mg, whose half-lives are 9.46 min and 21.1 h, respectively. There are several ways to produce ^{27}Mg, one of which is by the nuclear reaction of ^{26}Mg$(n, \gamma)^{27}$Mg. However, the target nuclide and the produced radioactive nuclide are the same element, which means that when ^{27}Mg is produced from ^{26}Mg, ^{27}Mg cannot be separated from ^{26}Mg, a stable nuclide. This means that the produced nuclide, ^{27}Mg, cannot be a carrier-free nuclide and always occurs together with an overwhelming amount of stable Mg elements. That is, when ^{27}Mg was employed as a tracer for the Mg experiment, because of the high Mg ratio to ^{27}Mg, a low concentration of Mg solution could not be prepared for the work: the radioactivity of ^{27}Mg in Mg solution is too low to be detected. Therefore, the element of the irradiating target and the element of the nuclide produced should be different to be able to prepare any concentration of Mg solution for the experiment.

Together with the consideration of the half-life, ^{28}Mg was chosen, and the reaction ^{27}Al$(\alpha, 3p)^{28}$Mg was employed for production. The key was the chemical purification of the trace amount of ^{28}Mg produced in the target, which means the separation of ^{28}Mg from the macroscopic amount of Al. The chemistry of Al is rather difficult, and the brief preparation is as follows.

Ten pieces of pure aluminum foil (99.999%, $10 \times 10 \times 0.1$ mm, each) were irradiated with a 50–75 MeV He(α) beam for 4–6 h by an AVF cyclotron installed at Tohoku University or QST (National Institutes for Quantum and Radiological Science and Technology) in Japan. Then, the irradiated Al foils in the vessel were removed and dissolved with 3 M HCl. After drying, the residue was dissolved with 2 M NH$_4$SCN and passed through a Sep-Pak Plus tC18 column (Environment, Waters). The eluted solution was applied to a column filled with AG50W-X4 resin (H$^+$ form, 100–200 mesh, Bio-Rad), and ^{28}Mg was retained by the resin. After washing with 0.5 M oxalic acid and 0.01 N HCl, ^{28}Mg was eluted with 2 M HCl. Then, the eluted solution was dried, and ^{28}Mg was dissolved in pure water. In the present study, the ^{28}Mg acquired through this chemical procedure was approximately 4–5 MBq per preparation.

The basic Mg uptake behavior in a rice plant was studied by applying ^{28}Mg as a tracer, and some of the results are introduced below [5–7, 10, 12].

3.2.2   Mg Uptake Activity Using ^{28}Mg as a Tracer

Using ^{28}Mg as a tracer, the Mg uptake activity of the rice root was studied. The Mg uptake activity changed rapidly with Mg concentration. In the case of an Arabidopsis seedling, high- and low-affinity transport systems were revealed by a kinetic study, and the high-affinity transport system was upregulated by Mg-deficiency treatment, similar to that applied to a rice seedling. The increased absorption activity at lower Mg concentrations and the rapid response to Mg concentrations suggested an active Mg absorption mechanism in roots. The Mg^{2+} uptake system in roots was upregulated within 1 h in response to the low Mg^{2+} condition. However, the Mg deficiency-induced Mg^{2+} uptake system was shut down within 5 min when Mg^{2+} was resupplied to the environment (data not shown).

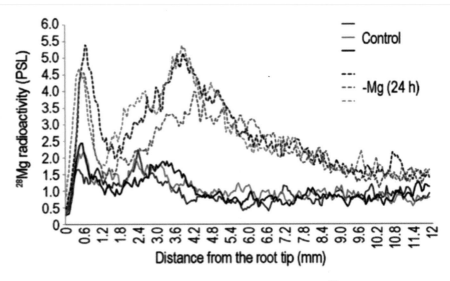

Fig. 3.21 Mg concentration in a root of Arabidopsis [6]. Distribution of ^{28}Mg along the main root of the control (1500 µM Mg^{2+} treatment for 24 h) and Mg plant (7 µM Mg^{2+} treatment for 24 h) after 30 min of ^{28}Mg absorption in the medium with 7 µM Mg^{2+}. Data from 3 control plants and 3 −Mg plants are presented

Most of the ^{28}Mg absorbed during 30 min was retained within the root tissue, and only a small percentage of ^{28}Mg was transported to the shoot. Magnesium in roots was not evenly distributed in the main root. The ^{28}Mg accumulation amount in the area between 2.4 mm and 6.0 mm of the Arabidopsis root tip was increased more than threefold under Mg-deficient treatment (Fig. 3.21), suggesting a high probability that Mg^{2+} is chiefly upregulated in this area.

To determine the movement of Mg after absorption by different parts of the root, a compartment box was prepared, which separated the root region every 1 cm to supply ^{28}Mg to a particular root region (Fig. 3.22). When Mg was applied to the upper part of the root, R-C, approximately half the Mg was transferred downward toward the root tip, whereas less than 5% of Mg moved downward when supplied to the middle part of the root. Figure 3.23 summarizes the orientation of ^{28}Mg movement, outside the originally supplied part of the root, when supplied from different parts of the root. It was shown that ^{28}Mg was already found in the crown roots after 15 min of ^{28}Mg treatment, although ^{28}Mg was hardly detected in the shoot. Then, the percentage of translocation to the shoot increased with time. When ^{28}Mg was supplied from R-C, the upper part of the root, the percentage of ^{28}Mg allocated to the lower root part, which consisted of R-A and R-B, was more than 50%, and only a small portion of the ^{28}Mg absorbed from R-B was detected in the lower root, R-A.

This result that a relatively large amount of Mg was transported downward from R-C was of particular interest. The pulse chase experiment clearly demonstrated that Mg absorbed from R-C was transported toward the root tip area without going through the upper root part or shoot (data not shown). The results suggested that

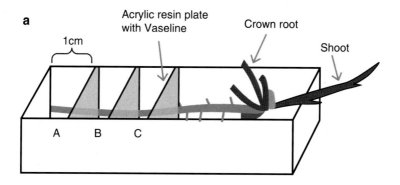

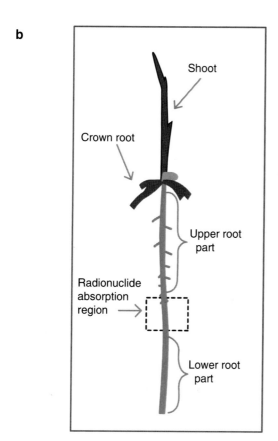

Fig. 3.22 Schematic illustration of a multi-compartment transport box [7]. (**a**) Schematic illustration of the multi-compartment transport box. Four or five rice (*Oryza sativa* L. cv. Nipponbare) seedlings were lined at the bottom of the box (only one seedling is shown as an example); then, acrylic resin plates were put through the grooves to partition each 1-cm-long compartment. The interstices were sealed with Vaseline to prevent leakage of solution from each compartment. The root regions in compartments A, B, and C were defined as R-A, R-B, and R-C, respectively. (**b**) Definition of the sample part. When the radionuclide was absorbed from R-A, the sample "lower root part" did not exist

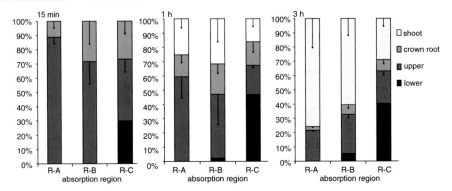

Fig. 3.23 Orientation of the [28]Mg movement when supplied from a specific region of rice roots [7]. Distribution of [28]Mg transported from each absorption region during 15 min, 1 h, and 3 h. The relative amounts of [28]Mg in the rice shoot (*white*), crown root (*light gray*), upper root part (*gray*), and lower root part (*dark gray*) are expressed as percentages. Mean standard deviations are presented ($n > 4$). When [28]Mg was absorbed from R-A, the data for the "lower root part" did not exist)

phloem loading of Mg occurred vigorously in the lateral root developing zone within minutes after uptake from the culture solution.

This characteristic movement of Mg according to the absorption site of the root was not observed for ^{32}P-phosphate or [45]Ca, as the upper part of the root did not show high transport activity, and R-C and R-B showed similar transport behavior (data not shown). These results indicated that the long-distance transport of Mg was controlled by a different mechanism from that of phosphate and Ca.

These findings led to our performing a study to determine the predominant response to Mg deficiency in rice seedlings. Each leaf was analyzed in terms of chlorophyll, starch, anthocyanin, and carbohydrate metabolites, and among several mineral deficiencies, only the absence of Mg was found to cause irreversible senescence of the fifth leaf (L5). The results suggested that the predominant response to Mg deficiency is a defect in transpiration flow. Furthermore, changes in myo-inositol and citrate concentrations were detected only in L5 when transpiration decreased, suggesting that they may constitute new biological markers of Mg deficiency (data not shown).

The movement of an element results from the combination or balance between xylem and phloem flow and is very complicated, similar to water flow [11]. Therefore, the only method to solve these problems in the dynamic movement of elements is the application of radioisotopes for tracer work.

3.2.3   Radioactive Tracer Production of K

Potassium (K) is one of the major essential elements in plants, and its physiological role and the role of transporters mediating transmembrane K transport within plant

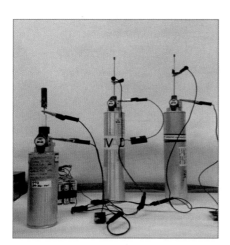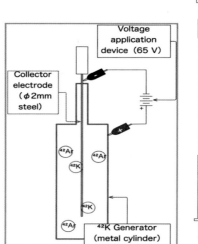

Fig. 3.24 ^{42}Ar-^{42}K generator. ^{42}K (half-life: 12 h) was prepared from the ^{42}Ar-^{42}K generator, where ^{42}Ar (half-life: 32.9 years) gas was sealed in a cylinder. The electrode was inserted in the cylinder, and 65 V was applied. After 3–4 days, the electrode was removed and washed in a water solution at 42 K, and the decay product of ^{42}Ar collected to the electrode was dissolved as a carrier-free ^{42}K^{+} ion

tissue have been intensively studied. Despite these studies, the behavior of long-distance K transport and the factors affecting K, as well as the distribution and function of the transporters, have not yet been clarified. To understand K behavior in a living plant, it is necessary to develop a new technique using a radioactive nuclide of K. There are two radioactive nuclides as candidates to trace K behavior, ^{38}K and ^{42}K, with half-lives of 7.6 min and 12.4 h, respectively. Because of their short half-lives, neither of the nuclides is commercially available.

^{38}K can be produced by the ^{38}Ar(p, n)^{38}K reaction using a small cyclotron. Although we once employed ^{38}K as a tracer to study K uptake in a rice root (See previous section), only a short experiment, within approximately an hour, was possible because of the extremely short half-life. Therefore, the other radioactive nuclide, ^{42}K, is preferable to trace a longer time of K movement. The device to produce ^{42}K is as follows.

^{42}K can be prepared from a ^{42}Ar-^{42}K generator (Fig. 3.24) [13]. The generator is filled with ^{42}Ar gas, which decays constantly, producing ^{42}K gas in the container. ^{42}Ar has a half-life of 32.9 years and can be produced through the ^{40}Ar(t, p) ^{42}Ar reaction by irradiating Ar gas, which contains 99.6% stable ^{40}Ar, with a tritium (^{3}H) beam using a cyclotron. The produced ^{42}Ar gas was transferred into a steel cylindrical container to generate ^{42}K. Inside the cylinder, ^{42}Ar decayed constantly, according to its half-life, and produced ^{42}K gas. To collect the ^{42}K gas produced in the cylinder, a steel cathode was inserted, and approximately 60 V was applied so that the $^{42}K^+$ gas adsorbed on the steel cathode. Since the half-life of ^{42}K is 12.4 h, after approximately 2 days (after 4 half-lives), the amount of ^{42}K produced was 92% of the maximum amount produced at equilibrium. Therefore, a few days were needed to collect $^{42}K^+$ on the cathode. Then, this cathode with adsorbed carrier-free ^{42}K was washed for a few minutes with water in a glass tube containing a low concentration of KCl to obtain a ^{42}K solution. Approximately 5 KBq of ^{42}K was obtained at each collection. Because of the lack of a radioactive tracer of K, ^{86}Rb has sometimes been used as a substitute; however, there is no evidence that ^{86}Rb undertakes the physiological role of K or completely traces the behavior of K.

The greatest advantage of using the ^{42}Ar-^{42}K generator is that ^{42}K can be prepared repeatedly in the laboratory. Using the ^{42}K tracer, real-time imaging of the element movement in a plant was performed, and the results are shown in Chap. 4.

3.3   Other Elements

Finally, when considering the elements in plants, it should be noted that naturally grown plants have different concentrations of elements according to the site where they grow. Plants adapted to the nature of the soil have been selected or acquired specific physiological features to survive through the long history of evolution. Sometimes, at high concentrations of heavy elements, plants adapt to live without showing any effect from the poisonous element. For example, some *Astragalus* sp., a grass that grows in meadows, has adapted to grow at high soil concentrations of Se and accumulates high amounts of Se, sometimes 1000 times higher than those in other plants, thereby becoming a poisonous plant. Se toxicity can kill a mouse that is fed wheat grown in soil containing 1 mg of Se per 1 kg. When animals eat this grass, S in the chemical structure of two essential amino acids is replaced with Se, causing serious disease in animals. The deaths of sheep or horses caused by this poisonous plant were described in Marco Polo's diary in the thirteenth century. This grass favors Se and acquired a new metabolic pathway to escape the poisonous effect of Se. In addition to this example, the profiles of elements in naturally grown plants could provide information about the features of the area and what elements are present in high quantities. Considering the adaptation of plants to areas with different kinds and concentrations of the elements in the soil, the elements in a plant could be described as another DNA, reflecting the environmental history of the growth site.

3.3.1   Production Districts of Onion and Beef

Since plant growth is highly dependent on the elemental concentrations in soil, the elemental profile of a plant is expected to be correlated well with that of the soil. This

Fig. 3.25 Element profiles of onions produced in different districts [8]. Elements in onion cultivars harvested from 14 and 20 points of Hokkaido and Saga, respectively, were analyzed by neutron activation analysis. The B, S, and Cl concentrations were determined by a prompt γ-ray analysis. In particular, Cl was an important element in the production districts of onions

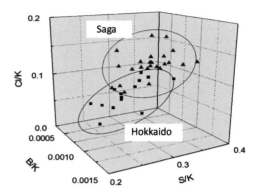

means that the amount or profile of the elements contained in a plant could be an indicator of where the plant was grown.

Recently, consumers have paid close attention to the quality of agricultural products, especially the production districts, to verify the safety of the products. For the same cultivar, there is no difference in the DNA sequence of plants growing in different districts. However, the same kind of plant must have different elemental profiles when produced in different districts.

To determine the elemental concentration in plant samples, inductively coupled plasma-mass spectrometry (ICP-MS) is now widely performed, which requires acid digestion of the sample. However, some volatile elements, such as I and Cl, are lost during the sample preparation process. To avoid this problem, it is preferable to apply a nondestructive analytical method that does not require acid digestion of the sample. Therefore, neutron activation analysis (NAA) and prompt gamma-ray analysis (PGA) methods were employed to determine the amounts of different elements, including elements that cannot be determined from the method using acid-digested samples. These methods are the best tools for this kind of analysis since they allow nondestructive multielement determination.

Considering the agricultural products from different districts, not only plants but also animals could be candidates to analyze the growth site. For example, cows feed on haylage, fermented local pasture that might contain the specific elemental profile of the site. As an example of using this kind of analysis to search for differences in production district among the agricultural products, two cases are presented below, onion and beef.

Onions grown in the northern part of Japan (Hokkaido) and the southern part of Japan (Saga) were selected. Approximately 500 mg of each sample was dried, sealed doubly in a well-washed polyethylene vinyl bag, and irradiated in the JRR-3 M research reactor installed at the Japan Atomic Energy Agency in the same way described above. NAA was used to determine Na, Mg, Cl, K, Ca, and Mn, and PGA was used to determine H, B, C, S, Cl, and K by γ-ray spectroscopy. The data set obtained by NAA and PGA was analyzed using principal component analysis

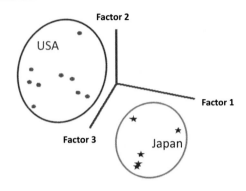

Fig. 3.26 Comparison of elemental profiles in beef produced in the USA and Japan [9]. Ten elements of Japanese black cattle beef produced in the USA were analyzed by neutron activation analysis and prompt γ-ray analysis. These beefs were sufficiently grouped by principal component analysis (PCA) with the elemental data set

(PCA). The onion production site could be distinguished using 7 elements, B, S, Cl, Na, Mg, K, and Ca, according to PCA using Pirouette application software (vr. 3.11, Informetrix). Figure 3.25 shows one of the results categorizing the production site, obtained from 3 sets of elemental ratios: Cl/K, B/K, and S/K. The production site of the onion could be clearly distinguished based on the data set of the elements. Although Cl is a difficult element to measure through the digestion method, the Cl concentration in onion was found to be prominent among the elements measured to identify the production districts of the onions.

Beef samples produced in different districts in Japan, Australia, and the USA were analyzed by NAA and PGA. Freeze-dried samples of chuck, sirloin, fillet, round and other parts were prepared for PGA analysis, and the elements were analyzed in the same way as those in onions. NAA was used to determine Na, Na, Mg, Cl, K, Br, and Sm, and PGA was used to determine H, C, N, and S by γ-ray spectroscopy. With the same data processing used for onions, PCA, it was possible to group Japanese black cattle beef separately from beef from the USA.

However, Holstein beef from Japan and from Australia were not sufficiently grouped by PCA modeling with the elemental data set. It was also found that there was no difference among the parts of the beef, round, sirloin, chuck, and fillet in the grouping of production districts by PCA modeling. These are the first examples of identifying the production site of onions and the provenance of beef through elemental analysis.

3.3.2 Other Elements

Since plants grown in different districts could reflect the mineral composition or concentration of the soil, the author tried to analyze gold (Au) in plants to search for gold mines. The sensitivity of NAA to Au is extremely high, and trace amounts of ^{198}Au can be produced in a reactor by the (n, γ) reaction (Table 3.1) and measured by γ-spectroscopy. This high sensitivity is beyond what people normally estimate; as cited above, when gold accessories or wristwatches are worn during the sample preparation, the gold vapor from the gold materials contaminates the sample and can be detected in the measurement. Many kinds of plants and the corresponding soils were collected at an interval of a certain distance from the gold mine. Then, the plant samples were washed well, and the Au amount in both plants and soils was measured by NAA. The Au amount in the plant increased at a greater distance from the mine than the amount in the soil. Analysis of the plants grown around the mine showed that *Callicarpa mollis*, beauty-berry, accumulated high amounts of gold and was promising as an indicator for the gold mine; however, the gold amount in the plant was on the order of 10^{-9} g.

3.4 Summary and Further Discussion

The application of NAA into plant samples and the results are presented. NAA has been performed in elemental analysis for many years, and there has been remarkable progress in NAA technology, such as PGA, which enabled the nondestructive measurement of H, B, or halogens. The application of TOF (time of flight) technology in the measurement is another developing technique, taking advantage of analyzing time-dependent γ-ray spectroscopy. The overall features of NAA are as follows.

1. NAA is the only method to determine the absolute amounts of elements in both solid and liquid samples.
2. The sensitivity of NAA is extremely high, especially the sensitivity to heavy elements, although the sensitivity differs among elements.
3. It allows multielement analysis at one time, and a time-dependent measurement is needed to determine each element.

The multielement analysis of plants provided much information. Within a plant, element-specific concentrations were found to spread throughout the whole plant tissue. Each element formed a specific gradient or illustrated barriers between the tissues, which were suggested to regulate the movement of the elements. The barrier between roots and aboveground parts is well known and is a common feature of heavy elements. Element barriers were found even in the same root or in the same tissue of the aboveground part, showing imbalance of the element profile. Although the element profile of the elements showed a similar tendency during the developmental stage, changes in the profile were observed between the juvenile and senescent stages.

When there is a change in the element profile, the need for element-specific movement can be estimated, and there is a specific requirement time for each element. Then, why do the elements move? They might respond to growth and environmental conditions, especially light, which could feature plant activity. Then, what about the roots? There were complicated routes for element absorption and movement within a root in the dark, depending on the position. The root tip showed its specific feature for element accumulation and movement. It was interesting to know the Al movement at the root tip. Though the individual tissues showed different concentration profiles of the elements, the multielement movement in a whole plant should always be taken into account because every tissue is connected with each other.

In addition to the element profiles in the plant, the concentration of the elements in a plant reflected the mineral composition or concentration of the soil, and the amount of the elements contained in a plant could be one of the indicators of the site where the plant was grown.

Radioactive nuclides are indispensable tools for tracer work in plants. However, when a suitable RI is not commercially available, the method of production should be considered. In this sense, the production of ^{28}Mg and ^{42}K was presented. Since ^{28}Mg was not previously used for plant research, the physiological function of Mg could be shown in more detail using ^{28}Mg as a tracer. Although further experiments using ^{28}Mg were omitted in this book, some experiments are presented to show that the physiological study of the plant is highly dependent on radioactive tracers.

Bibliography

1. Nakanishi TM, Matsumoto S (1992) Element distribution in barley during the developmental stage. Radioisotopes 41:64–70
2. Nakanishi TM, Tamada M (1999) Kinetics of element profile pattern during life cycle stage of morning-glory. J Radioanal Nucl Chem 239:489–493
3. Nakanishi TM, Furukawa J, Ueoka S (2000) Kinetics of transition element profile during the life cycle of morning-glory. J Radioanal Nucl Chem 244:289–293
4. Ikeue N, Hayashi Y, Tanoi K, Yokota H, Furukawa J, Nakanishi TM (2001) Circadian rhythms in the concentration of Mg, Ca and Al in *Pharbitis nil*, analyzed by neutron activation analysis. Anal Sci 17(supplement):1491–1493
5. Kobayashi N, Tanoi K, Nakanishi TM (2006) Magnesium localization in shoot apices during flower induction in *Pharbitis nil*. Canadian J Botany 84:1908–1916
6. Ogura T, Kobayashi NI, Suzuki H, Iwata R, Nakanishi TM, Tanoi K (2018) Magnesium uptake characteristics in Arabidopsis revealed by ^{28}Mg tracer studies. Planta 248:745–750
7. Kobayashi NI, Iwata N, Saito T, Suzuki H, Iwata R, Tanoi K, Nakanishi TM (2013) Different magnesium uptake and transport activity along the rice foot axis revealed by ^{28}Mg tracer experiments. Soil Sci Plant Nutr 59:149–155
8. Tanoi K, Matsue H, Iikura H, Saito T, Hayashi Y, Hamada Y, Nishiyama H, Kobayashi N, Nakanishi TM (2008) Element profiles of onion producing districts in Japan, as determined using INAA and PGA. J Radioanal Nucl Chem 278:375–379
9. Saito T, Tanoi K, Matsue H, Iikura H, Hamada Y, Seyama S, Masuda S, Nakanishi TM (2008) Application of prompt gamma-ray analysis and instrumental neutron activation analysis to identify the beef production distinct. J Radioanal Nucl Chem 278:409–413
10. Kobayashi NI, Saito T, Iwata N, Ohmae Y, Iwata R, Tanoi K, Nakanishi TM (2013) Leaf senescence in rice due to magnesium deficiency mediated defect in transpiration rate before sugar accumulation and chlorosis. Physiol Plant 148:490–501
11. Kobayashi NI, Tanoi K, Hirose A, Nakanishi TM (2013) Characterization of rapid intervascular transport of cadmium in rice stem by radioisotope imaging. J Exp Bot 64:507–517

12. Tanoi K, Kobayashi NI, Saito T, Iwata N, Hirose A, Ohmae Y, Iwata R, Suzuki H, Nakanishi TM (2013) Application of ^{28}Mg to the kinetic study of Mg uptake by rice plants. J Radioanal Nucl Chem 296:749–751
13. Aramaki T, Sugita R, Hirose A, Kobayashi NI, Tanoi K, Nakanishi TM (2015) Application of ^{42}K to Arabidopsis tissues using Real-Time Radioisotope Imaging System(RRIS). Radioisotopes 64:169–176

Chapter 4
Real-Time Element Movement in a Plant

Keywords Real-time RI imaging · Real-time RI imaging system · Microscopic RI imaging system · Rice · Soybean · Arabidopsis · ^{14}C · ^{22}Na · ^{28}Mg · ^{32}P · ^{33}P · ^{35}S · ^{42}K · ^{45}Ca · ^{54}Mn · ^{65}Zn · ^{109}Cd · ^{137}Cs · Image analysis · Root absorption image · Micro-movement · Imaging system development

4.1 Conventional Radioisotope (RI) Imaging

Radioisotope (RI) imaging has been developed as radiography using X-ray film. RI was supplied to the plant, and the plant was placed on an X-ray film for exposure to acquire the radiation image on the film. With further development of the technology for this kind of radiography, an IP (imaging plate) replaced the X-ray film, providing much higher sensitivity. Since the image by an IP showed a linear relationship between the whiteness in the image and the radiation intensity, it was easy to quantify the amount of the radionuclide from the image that appeared on the IP. With the development of gene technology, where ^{32}P was employed to label DNA, IP was widely used to detect the RI band of DNA, especially in electrophoresis.

Figure 4.1 shows how to obtain an RI image of the plant sample. After treatment with RI, the plant was placed in a cassette containing an X-ray film or an IP, and the film or IP was exposed to the radiation from the sample for a time. After exposure, the X-ray film was developed, and the image was acquired by a scanner. In the case of an IP, the image produced in the IP was acquired by an image scanner and stored in a computer. After scanning, the image in the IP could be erased, and the IP could be used repeatedly for other samples.

Figure 4.2 shows an example of an RI radiograph of a soybean plant 2, 4, and 6 days after pulse (2 h) treatment of the root with 0.325 kBq/mL of ^{109}Cd solution. Figure 4.2a is the image of the whole plant showing the distribution of ^{109}Cd. Figure 4.2b shows the dissection images of the plant taken by an IP, suggesting

Supplementary Information The online version of this chapter (https://doi.org/10.1007/978-981-33-4992-6_4) contains supplementary material, which is available to authorized users.

T. M. Nakanishi, *Novel Plant Imaging and Analysis*,
https://doi.org/10.1007/978-981-33-4992-6_4

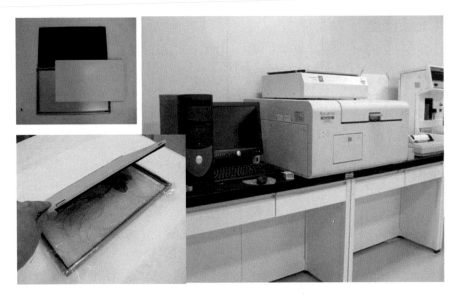

Fig. 4.1 RI (radioisotope) imaging system using an IP. A plant sample containing RI is placed on an IP shown as a white board, and the IP was exposed to radiation from the sample for a time period in a cassette. Then, the radiation image is scanned by a scanner, and the RI amount is analyzed by a computer

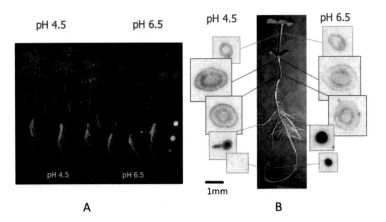

Fig. 4.2 An example of a ^{109}Cd image in a soybean plant. ^{109}Cd distribution in the whole plant (**a**) and the dissection (**b**) when ^{109}Cd was supplied under different pH conditions. (**a**): pseudocolor was added according to the intensity of the radiation; (**b**): darker colors indicate higher radioactivity

that Cd translocation was performed mainly via the vascular bundle. The resolution of the image obtained using an X-ray film was approximately 20 µm, and that obtained using an IP was approximately 50–100 µm.

These are the widely applied methods of RI radiography. To maximize the resolution and contrast of the image, the sample should be kept as close as possible to the film or the IP during the exposure. A cassette is usually used for exposure of

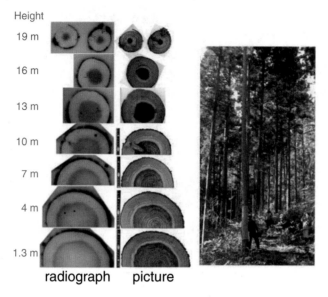

Height

19 m

16 m

13 m

10 m

7 m

4 m

1.3 m

radiograph   picture

Fig. 4.3 Trunk cross section image of ^{137}Cs [1]. One year after the Fukushima nuclear accident, a cedar (*Cryptomeria japonica*) tree grown in the mountain in Fukushima, where the environmental radioactivity was approximately 0.2 μSv/h, was downed, and wood disks (5-mm thickness) were removed from the same log. The ^{137}Cs distribution image in the disk was acquired using an IP (BAS-IP MS, GE Healthcare Japan) after 1–5 months of exposure time. The image in the IP was scanned by a fluorescent image analyzer (FLA-9000, Fujifilm)

the film or IP. However, if a cassette is used, the same plant cannot be used for further experiments after being kept in the cassette, since the plant was pressed tightly by the cassette lid for a while.

As another example, a radiograph of a wood disk is presented, which was taken after the Fukushima nuclear accident (Fig. 4.3). The cedar tree was cut down; wood disks, 1 cm in thickness, were removed from different heights of the tree; and the radiation image of ^{137}Cs was taken by an IP. The radiographs showed that fallout ^{137}Cs accumulated at the trunk surface and heartwood, especially at the higher positions in the tree.

These are static images of harvested plant samples, taken by an IP. However, to obtain images of living plants, several devices are needed. For example, to determine the RI distribution in roots, the following device was prepared. A container for culture solution was employed in which plant roots treated with RI were pushed down gently with a mesh flange to bring all the root tissue into contact at the bottom. From outside the bottom of the container, an IP was placed for a time to take the root image. The IP could be replaced periodically to acquire successive RI distribution images of the root. Figure 4.4 is an example of a rice plant where 0.4 GBq/50 mL of ^{38}K (half-life: 7.6 min) was supplied to the plant. An IP was placed outside the container for 20 min to acquire the root image from the tip to the upper part. Tracing the root image allowed the amount of ^{38}K in the root to be calculated from the darkness of the image.

Fig. 4.4 An example of taking an image of ^{38}K absorption of rice roots. The rice roots supplied with ^{38}K in the culture solution were gently downed with a mesh board to the bottom of the container. An IP was attached at the outside of the bottom part of the container to acquire the radiation image of the roots

Similar usage of an IP was introduced in Part I, Chap. 2, Sects. 2.3.1 and 2.3.2, where an IP was placed close to a plant supplied with ^{15}O-water, and by changing the IP, another image of ^{15}O-water distribution was acquired. The difference between the images showed water movement.

4.2 Development of a Macroscopic Real-Time RI Imaging System (RRIS)

4.2.1 Construction of RRIS (First Generation)

Using an IP, live images of a living plant could be acquired by changing the IP successively; however, adjusting the positions of the images in different IPs was difficult, and it was not possible to trace the fast movement of elements. Therefore, a real-time imaging system was developed that allowed successive images to be taken and allowed the imaging of commercially available, conventional RIs. First, we adjusted all the devices with ^{32}P, whose β-ray energy is relatively high (1709 keV), offering the advantage of high-efficiency scintillator conversion. From the plant physiological point of view, phosphate is an important component of nucleic acids, and phospholipids play an important role in energy transformation. It is also known

Fig. 4.5 Principle of the real-time RI imaging system [2]. The radiation emitted from the sample was converted to light by a scintillator deposited on a fiber optic plate (FOP), and the light was detected by a highly sensitive single-photon counting camera to produce a radioactivity image

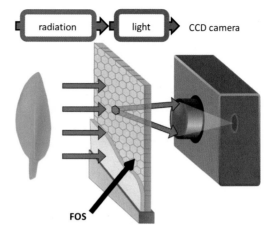

(Fiber optic plate with CsI(Tl) Scintillator)

that phosphate supports photosynthesis as a substrate for the reaction and mediates signal transmission, etc.

The system we developed consisted of the following two steps: (1) conversion of β-rays emitted from plants into light by a scintillator and (2) detection of the light by a highly sensitive, single-photon-counting camera. The details of the imaging process were as follows. The β-rays from RI were converted to light by a scintillator; however, the intensity of the light was very low, and therefore, amplification of the light signal was needed. The light was amplified by an image intensifier unit (with GaAsP semiconductors in a photocathode), where light was converted to electrons in the photocathode, followed by amplification within a microchannel plate (MCP). The MCP consists of many thin channels of glass (capillary) to amplify electrons. When electrons were accelerated in an electric field by high voltage and introduced to capillaries, the electrons clash with the opposite side to emit more electrons. After several repetitions of these processes, an image was produced on the fluorescent surface, which was prepared on the MCP. The diameter of the capillary was 6 μmφ, which was the smallest size of MCP available now; however, this size was a key factor limiting the resolution level of the image. An image produced by electrons on a fluorescent surface was detected by a CCD camera in the AQUACOSMOS/VIM system (VIM system). Thirty image frames, consisting of 350,000 pixels/frame, were acquired per second, and the image was integrated for 1–3 min.

The area of the image covered 5.0 × 7.0 cm of the plant sample. The measurement had to be performed in a dark box, since the light intensity from the sample was very weak even for a highly sensitive cooled CCD camera for single photon counting. This type of highly sensitive camera is severely damaged when environmental light leaks into the dark box. Therefore, first, the whole imaging system including the sample was kept under completely dark conditions. The sample was kept horizontal to acquire the plant image in the first generation of the imaging

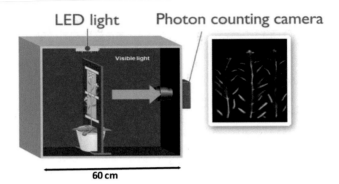

Fig. 4.6 Overall image of the real-time RI imaging system (first generation). To prevent the damage of the light to a CCD camera, all imaging processes were performed in a *dark box*

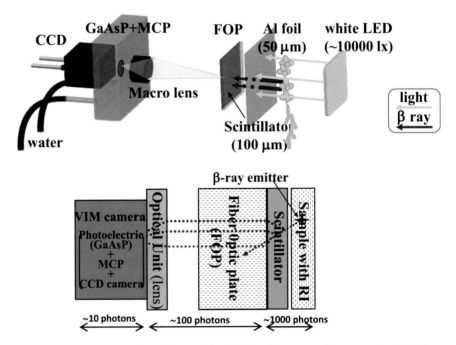

Fig. 4.7 Schematic illustration of the real-time RI imaging system. To prevent the light from irradiating the plant, the FOP was covered with an Al foil. The β-ray penetrating the Al foil was converted to light by a scintillator and amplified by a GaAsP intensifier. Then, the light taken by the CCD camera was accumulated and produced the radiation image

system (Fig. 4.5). In the second and third generation of the imaging system, the sample was always kept vertical during imaging (Figs. 4.6 and 4.7).

4.2.2   Performance of RRIS

4.2.2.1   Dynamic Range of the System

The important parameters for the performance of the imaging system are sensitivity, resolution, and quantitative treatment of the image. However, considering the RIs to be applied, the performances are dependent on the energy and kind of radiation emitted from each nuclide. Seven nuclides, ^{14}C, ^{22}Na, ^{28}Mg, ^{65}Zn, ^{86}Rb, ^{109}Ca, and ^{137}Cs, were chosen to study the performance of the RRIS. Table 4.1 shows the features of the representative RIs used in the experiment. In the case of ^{28}Mg (half-life: 20.9 h), a radioactive equilibrium is attained with the daughter nuclide ^{28}Al (half-life: 2.2 min), whose energy is higher than that of ^{28}Mg. Therefore, the major contribution of the β-ray energy to imaging is considered to be derived from ^{28}Al.

To evaluate the performance of RRIS, a standard solution was prepared. Two to 3 μL of each standard solution was mounted on a polyethylene terephthalate sheet and covered with a polyethylene sheet (10 μm in thickness) after the solution was completely dried. In the case of ^{14}C, polyphenylene sulfide (1.2 μm in thickness) was used to cover the sheet. Then, the image acquired by RRIS was compared with that of an IP.

Table 4.2 shows the dynamic range of quantitative analysis in both RRIS and IP imaging measurements, with the lower limit and upper limit of the radiation counting. As shown in the Table, these properties were highly dependent on the kind of nuclide. To explain the lower limit of detection, ^{28}Mg measurement can be described as follows, as an example. In the case of RRIS, the lower limit for quantitative counting did not change with increasing accumulation time. The camera used in RRIS detected 1 photon per frame in 1 pixel at a speed of 30 frames/s. However, because a certain amount of noise exists at each count, such as dark current, the signal-to-noise ratio (S/N) does not improve with increasing

Table 4.1 Feature of the applicable nuclides for the RRIS

Nuclide	Decay	Half-life	β-ray energy (keV) av.	β-ray energy (keV) max.	γ- or X-ray energy (keV)
C-14	β$^-$	5700 years	49.5	157	–
Na-22	β$^+$, EC	2.6 years	216	546	1275, 551 (annihilation)
Mg-28	β$^-$	20.9 h	152	860	1589
(Al-28)	β$^-$	2.2 months	1242	2863	1779
P-32	β$^-$	14.3 days	695	1711	–
S-35	β$^-$	87.5 days	48.7	167	–
Ca-45	β$^-$	163 days	77.2	257	–
Mn-54	EC	312 years	–	–	835. 5.37 (Cr-Kα)
Zn-65	β$^+$, EC	244 days	143	329	1116, 551 (annihilation)
Rb-86	β$^-$	18.6 days	668	1774	1077
Cd-109	EC	461 days	–	–	22 (ae-Kα), 88 ($^{109\ mA}$Ag)
Cs-137	β$^-$	30.2 years	514	1176	662 ($^{137\ m}$Ba)

Table 4.2 Dynamic range of RRIS for the quantitative analysis

	RRIS				IP			
	The lower limit (Bq/ mm^2)	The upper limit (Bq/ mm^2)	Dynamic range	R-squared	The lower limit (Bq/ mm^2)	The upper limit (Bq/ mm^2)	Dynamic range	R-squared
C-14 (min)								
3	4×10^0	2×10^3	3×10^2	0.9972	4×10^3	4×10^3	1×10^3	0.9999
5	2×10^0	2×10^3	1×10^3	0.9976	2×10^0	4×10^3	2×10^3	0.9997
10	2×10^0	2×10^3	1×10^3	0.9981	1×10^0	4×10^3	4×10^3	0.9996
15	2×10^0	2×10^3	1×10^3	0.9983	1×10^0	2×10^3	1×10^3	0.9951
Na-22 (min)								
3	3×10^{-1}	3×10^2	1×10^3	0.9973	5×10^{-1}	6×10^2	1×10^3	0.9989
5	3×10^{-1}	3×10^2	1×10^3	0.9972	5×10^{-1}	6×10^2	1×10^3	0.9982
10	3×10^{-1}	3×10^2	1×10^3	0.9972	3×10^{-1}	6×10^2	2×10^3	0.9972
15	3×10^{-1}	3×10^2	1×10^3	0.9972	3×10^{-1}	6×10^2	2×10^3	0.9998
Mg-28 (min)								
3	3×10^{-1}	1×10^2	5×10^2	0.9993	6×10^{-1}	3×10^2	5×10^2	0.9966
5	1×10^{-1}	1×10^2	1×10^3	0.9995	3×10^{-1}	3×10^2	1×10^3	0.9975
10	1×10^{-1}	1×10^2	1×10^3	0.9995	3×10^{-1}	3×10^2	1×10^3	0.9985
15	1×10^{-1}	1×10^2	1×10^3	0.9989	1×10^{-1}	3×10	2×10^3	0.9975
Zn-65 (min)								
3	2×10^1	4×10^3	3×10^2	0.9992	2×10^1	2×10^4	1×10^3	0.9996
5	9×10^0	4×10^3	5×10^2	0.9990	2×10^1	2×10^4	1×10^3	0.9986
10	9×10^0	4×10^3	5×10^2	0.9990	9×10^0	9×10^3	1×10^3	0.9990
15	4×10^0	4×10^3	1×10^3	0.9990	9×10^0	9×10^3	1×10^3	0.9990
Rb-86 (min)								
3	5×10^{-1}	3×10^2	5×10^2	0.9957	1×10^0	6×10^2	5×10^2	0.9994
5	3×10^{-1}	3×10^2	1×10^3	0.9959	5×10^{-1}	6×10^2	1×10^3	0.9996
10	3×10^{-1}	3×10^2	1×10^3	0.9956	3×10^{-1}	6×10^2	2×10^3	0.9996
15	1×10^{-1}	3×10^2	1×10^3	0.9960	3×10^{-1}	3×10^2	1×10^3	0.9968
Cd-109 (min)								
3	2×10^0	1×10^3	5×10^2	0.9999	4×10^0	2×10^3	5×10^2	0.9997
5	1×10^0	1×10^3	1×10^3	1.000	2×10^0	2×10^3	1×10^3	0.9997
10	5×10^{-1}	1×10^3	2×10^3	0.9999	2×10^0	1×10^3	5×10^2	0.9999
15	5×10^{-1}	1×10^3	2×10^3	0.9999	1×10^0	1×10^3	1×10^3	0.9984
Cs-137 (min)								
3	5×10^{-1}	3×10^2	5×10^2	0.9964	5×10^{-1}	1×10^2	2×10^3	0.9979
5	3×10^{-1}	3×10^2	1×10^3	0.9983	5×10^{-1}	3×10^2	5×10^2	0.9996
10	3×10^{-1}	3×10^2	1×10^3	0.9988	5×10^{-1}	6×10^2	1×10^3	0.9950
15	3×10^{-1}	3×10^2	1×10^3	0.9989	3×10^{-1}	3×10^2	1×10^3	0.9999

L.L. and U.L.: Lower limit and upper limit of the quantitative radiation counting [3] modified

accumulation time, which resulted in a plateau in the count after 5–15 min of accumulation (Fig. 4.8a, b). This tendency was observed for most of the other nuclides. In the case of an IP, the image intensity is dependent on the accumulation

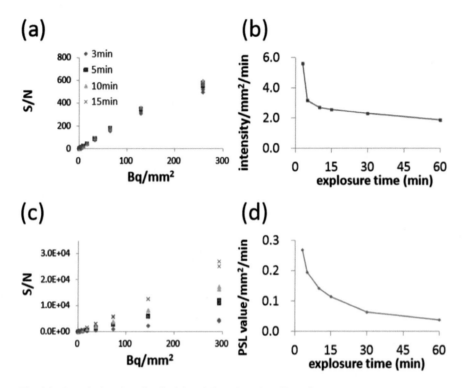

Fig. 4.8 Quantitative detection limit in RRIS and IP [3]. Effect of an accumulation time in RRIS and an exposure time in IP on the quantitative detection limit using the ^{28}Mg standard solution. (**a**) and (**b**) S/N ratio and detection limit for RRIS, respectively. Similarly, (**c**) and (**d**) are for an IP. The quantitative detection limit was calculated as $\sqrt{2} \times 10 \times \sigma$ (standard deviation of the background value)

of radiation energy in the photostimulable phosphor. Therefore, with increasing exposure time, the intensity of the image increased, and the S/N ratio improved (Fig. 4.8c, d).

When the upper limit of the counting is taken into account, the value in the RRIS did not change with the accumulation time, since there is an upper count limit per frame in the camera. For short-term counting, 3 or 5 min, there was no upper counting limit detected for the IP, suggesting that the upper counting limit was higher for an IP than for the RRIS. In contrast, for long-term accumulation, there was an upper limit for detection in an IP, suggesting that there was a limit to the amount of radiation that could accumulate in the photostimulable phosphor.

The results showed that the dynamic range of the RRIS was on the order of 10^3, suggesting that quantitative counting was possible even if the accumulation time was decreased from 15 to 3 or 5 min.

4.2.2.2   Distance Between FOS and the Plant

To acquire high image resolution, the distance between the plant sample and the FOS, on which the scintillator was deposited, should be as short as possible [4]. This distance will affect the quantitative analysis of the image. However, plants grow during imaging, which sometimes adds space between FOS and the sample. Therefore, evaluation should be performed to determine the relation between this distance and the intensity of the counting.

Six standard solutions of the nuclides ^{22}Na, ^{28}Mg, ^{65}Zn, ^{86}Rb, ^{109}Cd, and ^{137}Cs were prepared as point sources (3.3 mm^2), which were placed 0, 0.2, 0.3, and 0.4 mm from the surface of the FOS, and the radioactivity was counted. When the spot was 0.4 mm from the FOS surface, the counts decreased to 30–50% in all cases. One of the reasons for the decreasing count could be that the ROI (reason of interest) in the image was too small to cover the radiation, which was spatially expanded. Therefore, when the area of the ROI increased from 3.3 to 51 mm^2 (in the case of ^{28}Mg, from 5.3 to 46 mm^2), the counting was not dependent on the distance between the FOS and the standard sample (Fig. 4.9). The increased area of the ROI was set so that the neighboring standard did not cover each expansion of the radiation.

4.2.2.3   Self-Absorption

With a decrease in the β-ray energy, the self-absorption rate of the β-ray increases, which results in a decrease in the signal intensity in the image. Since the degree of self-absorption depends on the kind of tissue, tissues of Arabidopsis were used to measure the decrease in intensity. To investigate the self-absorption, 7 nuclides were applied to the Arabidopsis tissue. In the case of ^{14}C, ^{14}CO$_2$ produced by mixing ^{14}C-sodium hydrogen carbonate and lactic acid was supplied to the plant for 24 h. For the other nuclides, 3, 3, 10, 3, 10, and 0.5 kBq/mL of ^{22}Na, ^{28}Mg, ^{65}Zn, ^{86}Rb, ^{109}Cd, and ^{137}Cs, respectively, were supplied in culture solution and were grown for 3 days. After the treatment, flower parts, including the bulb and shoot tip, silique, stem, rosette leaf, and cauline leaf, were separated from the plant, and the intensity in the images acquired by the RRIS and radiation counts were compared. The radioactivity in each tissue was measured by a liquid scintillation counter for ^{14}C and by a γ-counter for the other nuclides.

Of the nuclides tested, the self-absorption effects for ^{28}Mg are shown as an example (Fig. 4.10). As shown in the figure, there was a good correlation between the image counts taken by RRIS and by a γ-counter in all the tissues investigated, where R2 was more than 0.9. Among the other nuclides, ^{22}Na, ^{65}Zn, ^{86}Rb, ^{109}Cd, and ^{137}Cs also showed good correlations (Table 4.3). Since the β-ray energy of ^{14}C is low compared to that of the other nuclides employed, the relative intensity of the radioactivity in the tissue decreased drastically with increasing thickness of the tissue, except in the thin tissues. Although the precise amount was unable to be calculated, the radioactivity of thin tissues such as flowers or rosette leaves was

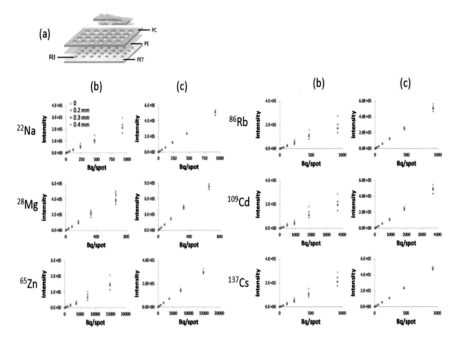

Fig. 4.9 Effect of the distance between FOS and a sample on the radiation measurement [3]. (**a**) Schematic illustration of the sample. Six standard radioisotope spots (^{22}Na, ^{28}Mg, ^{65}Zn, ^{86}Rb, ^{109}Cd, and ^{137}Cs) were prepared on a polyethylene terephthalate (PET) sheet and covered with a polyethylene (PE) sheet. On this sample, polycarbonate (PC) sheets (0.2, 0.3, and 0.4 mm in thickness) with holes were placed to maintain the distance between FOS and RI. (**b**) The ROI (region of interest) was set to be the same size as the standard RI spot (3.3 mm^2; except for ^{28}Mg: 5.3 mm^2). (**c**) The ROI was expanded to the maximum size to avoid superposing with the neighboring spots (51 mm^2; except for ^{28}Mg: 46 mm^2)

estimated with some errors permitted. When the correlation of ^{14}C was investigated, between the image counts acquired by RRIS and the radioactivity measurement by an IP, the result showed a high correlation, with an R^2 of 0.9857, suggesting that the contribution of the self-absorption rate of ^{14}C was the same for both measurements (Fig. 4.11). However, considering the low β-ray energy of ^{14}C, the performance of the RRIS should be taken into account carefully for ^{14}C imaging. This performance is described in more detail in the next Chapter, where the imaging of ^{14}CO$_2$ gas fixation is presented.

In all of the other cases of the elements tested, the correlation of the image between the RRIS and an IP was high. Figure 4.12 shows actual images of Arabidopsis tissues acquired both by the RRIS and by an IP. As shown in the figure, the two images are almost the same, suggesting a high correlation.

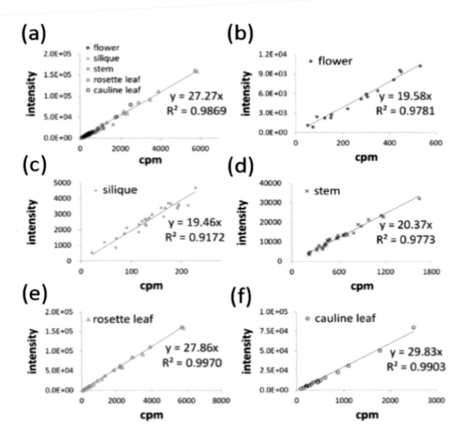

Fig. 4.10 Effect of self-absorption on the ^{28}Mg measurement [3]. The relation between the intensity of the image acquired by RRIS and the radioactivity measured by a γ-counter was plotted for the ^{28}Mg radioactivity absorbed in plant samples. (**a**) whole plant; (**b**) flower; (**c**) silique; (**d**) stem; (**e**) rosette leaf; (**f**) cauline leaf. The accumulation time for RRIS was 15 min, and the exposure time of the IP was 60 min

4.2.2.4   Simulation of Self-Absorption

Although the rate of self-absorption was found to be dependent on the kind and energy level of the radiation, there are several factors that induce self-absorption. One of them is the performance of FOS for each nuclide, since several kinds of radiation are emitted from each nuclide. To determine what kind of radiation contributes to self-absorption, a simulation was carried out. A Monte Carlo simulation code, EGS5, which is used in simulating medical exposure, was modified, and practical scintillator detection of FOS according to the kind of radiation was calculated. As shown in Table 4.4, the major contributions to the detection were β-rays, positrons, and X-rays, and the detection efficiency of γ-rays was small since γ-rays penetrate the scintillator.

Table 4.3 Correlation between the counting of the image taken by RRIS and radioactivity counting in tissues

	C-14	Na-22	Mg-28	Zn-65	Rb-86	Cd-109	Cs-137
Whole							
R^2	0.8396	0.9803	0.9869	0.9883	0.9659	0.9960	0.9937
Slope	1.006	10.10	27.27	2.342	83.77	2.083	5.579
n	111	90	109	76	90	101	117
Flower							
R^2	0.9815	0.9699	0.9781	0.9280	0.9196	0.9835	0.9269
Slope	0.6200	7.410	19.58	2.270	83.27	1.842	4.467
n	7	14	15	13	11	13	19
Silique							
R^2	0.7101	0.7088	0.9172	0.7504	0.8035	0.9863	0.9652
Slope	0.6830	6.898	19.46	2.014	68.80	1.888	4.695
n	14	24	27	24	24	24	39
Stem							
R^2	0.2263	0.9625	0.9773	0.9377	0.8535	0.8996	0.8249
Slope	0.8768	5.930	20.37	1.614	68.36	1.418	3.820
n	43	24	28	8	22	24	33
Rosette leaf							
R^2	0.8973	0.9964	0.9970	0.9932	0.9793	0.9971	0.9991
Slope	1.158	10.49	27.86	2.380	89.87	2.123	5.694
n	31	9	23	13	16	22	7
Cauline leaf							
R^2	0.6384	0.9520	0.9903	0.9638	0.9929	0.9887	0.9978
Slope	0.7808	9.458	29.83	2.308	80.39	1.987	5.602
n	16	19	16	18	17	18	19

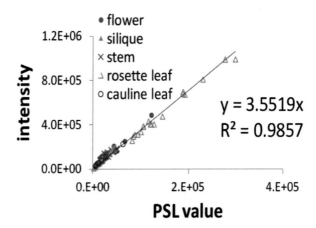

Fig. 4.11 Relation between the measurements of RRIS and IP [3]. After acquiring the image by RRIS (15 min), we measured the radioactivity of the standard sample prepared from ^{14}C-sucrose by a γ-counter for 2 min

$y = 3.5519x$

$R^2 = 0.9857$

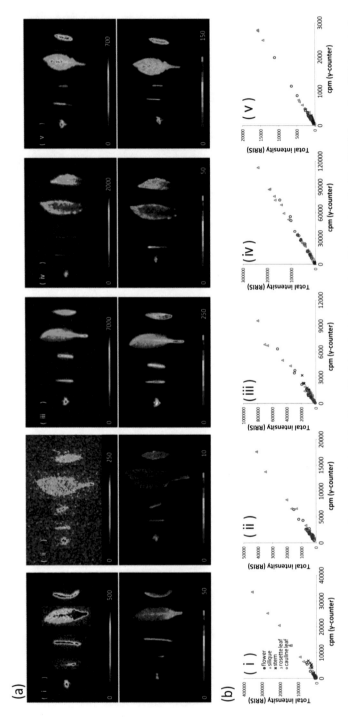

Fig. 4.12 Images of various *Arabidopsis* tissues containing RIs [4] modified. (**a**) Typical examples of tissue visualized by RRIS (upper) and IP (lower); from left to right: flower, silique, internode, rosette leaf, and cauline leaf. (i): ^{22}Na; (ii): ^{65}Zn; (iii): ^{86}Rb; (iv): ^{109}Cd; (v): ^{137}Cs. Pseudocolors were assigned according to the intensity/mm^2 for RRIS and PSL/mm^2 for an IP in the image. (**b**) Signal intensities detected in tissues were plotted against the radioactivity determined by the gamma-counting method

Table 4.4 Type of radiation detected by FOS by simulation [3]

	Radiation	Average kinetic energy (MeV)	Energy absorbed by the CsI (MeV/100,000decay)	Contribution percentage (%)
C-14	β	0.049	39.3	100.0
Na-22	Positron	0.216	1709.7	93.9
	7	1.275	72.4	4.0
	Ce Total	1.274	0.0	0.0
Mg-28	β	0.152	1004.0	64.0
	γ1	0.031	446.0	28.6
	γ7	1.373	39.7	2.6
	γ4	0.942	32.9	2.1
	γ2	0.401	31.4	2.0
	γ9	1.620	3.2	0.2
	γ8	1.589	2.5	0.0
	γ10	1.014	0.5	0.0
	Ce (K, γ1)	0.029	0.0	0.0
	Anger (K)	0.001	0.0	0.0
	Ce (L, γ1)	0.031	0.0	0.0
	γ3	0.607	0.0	0.0
	γ5	0.983	0.0	0.0
	γ6	1.342	0.0	0.0
	X-ray (K)	1.487	0.0	0.0
Al-28	β	1.242	5030.8	98.0
	γ	1.779	73.7	1.4
Zn-65	X-ray (Kα1)	0.008	44.3	38.9
	X-ray (Kα2)	0.008	22.2	18.4
	γ3	1.116	32.8	27.2
	X-ray (Kβ)	0.009	11.5	9.0
	Positron	0.143	9.0	7.5
	ce (K, γ3)	1.107	0.3	0.3
	Ce (L, γ3)	1.114	0.0	0.0
	γ2	0.771	0.0	0.0
	γ1	0.345	0.0	0.0
	Anger (K)	0.007	0.0	0.0
	X-ray (L)	0.001	0.0	0.0
	Anger (L)	0.001	0.0	0.0
Rb-86	β	0.668	4115.3	99.8
	γ	1.077	7.6	0.2
	X-ray (Kα1)	0.013	0.0	0.0
	X-ray (Kα2)	0.013	0.0	0.0
	Anger (K)	0.011	0.0	0.0
	Anger (L)	0.002	0.0	0.0
Cd-109	X-ray (Kα1)	0.022	394.4	48.4
	X-ray (Kα2)	0.022	210.4	25.8
	X-ray (Kβ)	0.025	131.0	16.1

(continued)

Table 4.4 (continued)

	Radiation	Average kinetic energy (MeV)	Energy absorbed by the CsI (MeV/100,000decay)	Contribution percentage (%)
	Ce (L)	0.084	29.5	2.6
	γ	0.088	27.9	3.4
	Ce (K)	0.063	8.0	1.0
	Ce (M)	0.087	6.4	0.8
	X-ray (L)	0.003	0.4	0.0
	Anger (K)	0.019	0.0	0.0
	Anger (L)	0.003	0.0	0.0
Cs-137	β	0.188	1472.2	69.2
	Ce (K, γ2)	0.624	443.6	20.8
	Ce (L, γ2)	0.656	77.3	3.6
	γ2	0.662	65.9	3.1
	X-ray (Kα1)	0.032	24.7	1.2
	Ce (M, γ2)	0.660	16.5	0.8
	X-ray (Kα2)	0.032	12.7	0.6
	X-ray (Kβ)	0.036	11.9	0.6
	Ce (N, γ2)	0.661	3.8	0.2
	X-ray (L)	0.004	0.2	0.0
	γ1	0.284	0.0	0.0
	Anger (K)	0.026	0.0	0.0
	Anger (L)	0.004	0.0	0.0

The other factor for self-absorption is the distribution of the nuclide within the tissue, which is not spatially uniform. To study the self-absorption derived from different distribution patterns of the nuclide within the plant tissue, a pipe model mimicking the stem was used, where the nuclide was distributed at the surface, interior, and center of the pipe (Fig. 4.13).

In the cases of ^{28}Al, ^{65}Zn, ^{86}Rb, and ^{109}Cd, there was no great difference in counting due to the distribution of the nuclide. In the cases of ^{22}Na, ^{28}Mg, and ^{137}Cs, the difference in the distribution affected the counting only of the nuclides distributed in the center. However, the vascular bundle, where nuclides are estimated to accumulate, is located relatively close to the surface, and it is not likely that the nuclides are present in the center of the stem. In the case of ^{14}C, there was a large difference in the effect of the distribution on the counting, especially between the homogeneous distribution and the localized distributions localized at the center or at the surface. Therefore, to apply ^{14}C for imaging, self-absorption must be taken into consideration. However, when the counting area was fixed and the change at the same site of the image with time was compared, numerical treatment of the change in the counting at this area was possible, as in the relative analysis.

The efficiency of producing the image was the integrated result of the various kinds of radiation emitted from the nuclide and its energy, specific to each nuclide.

Fig. 4.13 Simulation of the potential impact of steric distribution changes of RI [4]. To estimate the contribution of each radiation type from various RIs to RRIS imaging, cylindrical plant phantoms (2 mm in diameter) were prepared to perform the EGS5 simulation. The energy deposited in the Cs (Tl)I scintillator was calculated for each individual ray, and the total energy absorbed by the scintillator and contribution percentages of each radiation type were determined for each phantom. (**a**) Designed phantom; (**b**) imparted energy relative to ^{22}Na at the center

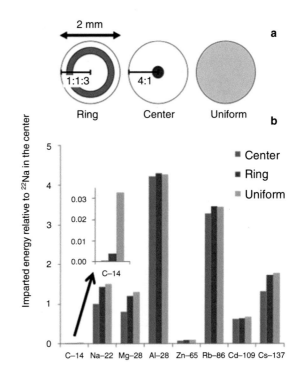

Generally, when the β-ray energy is high, the self-absorption is low, and the reverse effect occurs with low β-ray energy. In the case of X-rays, self-absorption is low, and the radiation penetrates the plant tissue but not the scintillator. High-energy γ-rays result in low counting efficiency since they penetrate both the plant tissue and the scintillator. With these results, it was possible to estimate the counting efficiency of other nuclides, for example, ^{32}P, which has a β-ray energy level similar to that of ^{86}Rb.

4.2.2.5 Actual Image of the Plant Sample

To acquire the plant image, a soybean plant (*Glycine max.* cv. Enley) was grown for 2 weeks, and then, Hoagland solution containing ^{32}P (P: 10 µM, 5 MBq/100 mL) was supplied for 24 h. Then, the ^{32}P image of the third trifoliate leaf was taken by RRIS. Figure 4.14 compares the images acquired by a prototype RRIS, as shown in Fig. 4.7, and an IP. It was found that the vein could be detected in a distinctly shorter time, a few seconds, than that needed for an IP, a few minutes.

Since the β-ray energy of ^{32}P is relatively high (max. energy: 1.7 MeV), another nuclide with lower β-ray energy, ^{45}Ca (max. energy: 0.257 MeV) was also used to compare the sensitivity of the imaging system. Figure 4.15 shows successive images of ^{45}Ca when this nuclide was supplied to the root of the soybean plant. The ^{45}Ca

Fig. 4.14 Comparison of the ^{32}P images acquired by RRIS and an IP [5]. [32]P images of a center leaf of the first trifoliate leaves of a soybean plant are shown after the treatment with ^{32}P for 24 h. Two images of the leaf at each imaging condition are shown. RI images acquired by the RRIS (*upper*) and an IP (*lower*) are presented. The ^{32}P image in a leaf can detect as early as 10 s by the RRIS

image in the third trifoliate leaf was taken after 1, 5, and 15 min by an IP (A) and after 10 s, 1 and 5 min by our imaging system (B). From the image series, it was found that image accumulation for 1 min by the RRIS provided a similar image to IP exposure for 15 min.

The results obtained by both nuclides indicated that the sensitivity of our imaging system was approximately 10 times higher than that of IP. The shorter accumulation

Fig. 4.15 Comparison of the [45]Ca images acquired by the RRIS and an IP [6]. The images of [45]Ca in trifoliate leaves by the RRIS (**a**) and IP (**b**). Veins are clearly observed in both imaging methods. The sensitivity of RRIS is higher than that of an IP, similar to those of ^{32}P images in Fig. 4.14

time for the RRIS indicates that a series of successive shorter images enables the production of a movie showing the real-time movement of the ions.

4.2.3 Imaging by Prototype Imaging System

4.2.3.1 [32]P Imaging in a Soybean Plant

Using the prototype imaging system, the behavior of ion uptake from the roots was shown. The first trial was ^{32}P-phosphate absorption in a soybean plant (*Glycine max.* cv. Enley). A soybean plant was grown for 2 and 6 weeks and then harvested to acquire leaf and young pod images, respectively. After [32]P (orthophosphate containing approximately 600 kBq/mL ^{32}P) solution was supplied from the root, several kinds of tissues from the whole plant were selected for imaging: meristem with young leaves, young trifoliate, center leaf of an expanded trifoliate, first leaf, and a young peapod.

Figure 4.16 is an image of the four tissues. In the leaves, it was expected that a high amount of ^{32}P-phosphate would be shown in the vein, since phosphate is transferred through the vein after absorption from roots. This was true in the first leaf. However, in contrast, high accumulation of phosphate was found between the veins in the first trifoliate leaf, suggesting a different route to transfer phosphate or leaking from the vein in the trifoliate leaf. A large amount of water was found to leak from xylem tissue in an internode (see Part I, Chap. 2, Sect. 2.3), and it was estimated to replace the water already present. In the case of phosphate, even if there was an accumulation of phosphate between the veins, there could not be any reason to

Fig. 4.16 ^{32}P-phosphate distribution in a soybean plant [7]. ^{32}P-phosphate was supplied from the root, and the accumulation pattern of ^{32}P was recorded by RRIS. ^{32}P-phosphate was first moved up to the youngest tissue and subsequently to the relatively older tissue. In one leaf, ^{32}P was found to be highly accumulated between the veins and shown as dots (*bottom left image*). The successive images with lap time are shown in Fig. 4.17

replace the phosphate already there. It was not known when and how phosphate leaked from the vein and why there was a different accumulation pattern between the first leaf and the first trifoliate leaf.

The successive images showed a rapid and high accumulation of ^{32}P in the youngest tissue, indicating that the absorbed phosphate was transferred primarily to the youngest tissue and then, when the ^{32}P in the youngest tissue was saturated, provided to the next younger tissues (Fig. 4.17). It seemed that ^{32}P-phosphate supply was prioritized according to the age of the tissue. This movement was obtained as a movie, and the lap time images are shown in Fig. 4.17.

Figure 4.18 shows how ^{32}P is transferred within the pod when supplied from the root. Since the image is based on radiation counting, the amount of ^{32}P-phosphate in the image could be analyzed. When the radioactivity in a specific area (marked as circle) was removed from the successive images, the behavior of the phosphate increase could be analyzed. Most of the ^{32}P-phosphate transferred in the direction of the pod initially accumulated at the bottom part of the pod during the first 2 h. Then, ^{32}P was transferred to the two seeds within the pod, not to the surface of the pod.

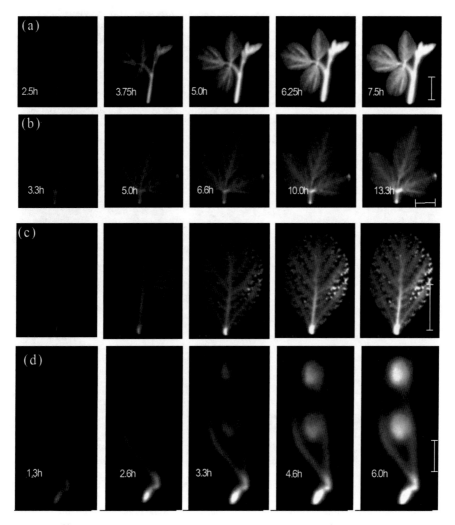

Fig. 4.17 ^{32}P translocation images in a soybean plant [8]. (**a**) meristem with young leaves; (**b**) young trifoliate; (**c**) center leaf of an expanded trifoliate; (**d**) young peapod. (**a**) 3-min integrated image, (**b**), (**c**), and (**d**) 4 min integrated image. *White bar*: 10 mm

When the accumulation of ^{32}P between the seeds was compared, the accumulation rate and amount in the two seeds were approximately the same.

4.2.3.2 ^{14}C Imaging in a Rice Plant

In this section, the visualization of ^{14}C-labeled amino acid absorption in a rice plant using the prototype RRIS is presented. The beginning of this research was as follows. Recently, organic farming, without depending on chemical fertilizer, has

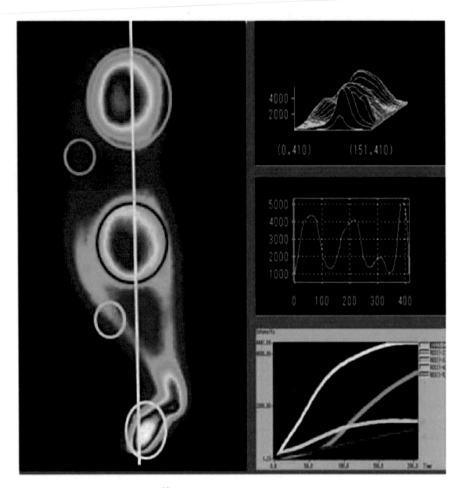

Fig. 4.18 Image analysis of the ^{32}P-phosphate accumulation in the pod [7]. The pseudocolor was assigned based on the counts in the pixels. First, ^{32}P accumulated at the bottom part of the pod and subsequently transferred to the other parts of the pod. There was no difference in the transfer speed or amount of ^{32}P between the two seeds

become popular, where incompletely decomposed organic matter, including plants, is supplied. However, it has been taken for granted that plants live on 17 inorganic ions and that other chemicals are not needed for growth. If plants can absorb amino acids, along with inorganic ions, this evidence could provide scientific support for organic farming from some perspectives. Therefore, the rice plant (*Oryza sativa* L. Nihonbare) was selected to investigate whether it could absorb amino acids and to analyze the chemical forms of the amino acids within the plant. The imaging part of this study, visualization of amino acid absorption, is shown below with some results obtained from chemical analysis.

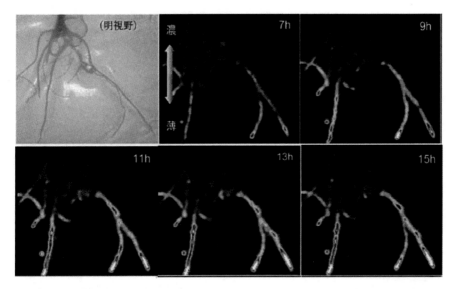

Fig. 4.19 ^{14}C-glutamine absorption images of the roots from the solution [9]. Twenty milliliters of ^{14}C-glutamine (18.5 kB/mL) solution were supplied to a two-day seedling of the rice plant, and the ^{14}C accumulation images of the root were monitored for 30 h. The accumulation time for each image was 10 min

First, by applying doubly labeled (^{15}N and ^{13}C) glutamine to 6-day-old seedlings, [15]N- and ^{13}C-glutamine were detected in both the underground and aboveground parts of the plant by a high-performance liquid chromatography equipped with an ion trap mass spectrometer. The results suggested that glutamine itself was absorbed from the root without decomposition in the rhizosphere and was transferred to the aboveground part.

Then, real-time imaging of glutamine absorption from roots was performed to analyze the uptake of amino acids in roots. The container with root was placed on a Cs(Tl)I scintillator that was covered with polystyrene film (4 μm in thickness). Then, by pushing the root gently with a sponge, the root was set close to the scintillator. Approximately 20 mL of ^{14}C-glutamine solution (18.5 kB/mL, containing 250 μM N) was supplied to the container, and the uptake of the amino acids in the roots was visualized (Fig. 4.19). The image was integrated every 10 min/frame until 43 h. When the successive images were produced as a movie, it was interesting to note that first, a ^{14}C-glutamine accumulation front appeared around the lower part of the roots in the solution, and then, the ^{14}C-glutamine abruptly moved upward and was absorbed by the roots, indicating that there was no continuous absorption of the amino acid uptake but suggesting that optimal timing and optimal concentration were needed for the ion movement toward the root in solution.

Within the same root, the uptake activity and accumulation of glutamine were not uniform, and high accumulation of glutamine was always shown at the root tip. The absorbed amount in the root plateaued after approximately 10 h. When transpiration was prevented by covering the leaf surface with Vaseline, the absorption rate and

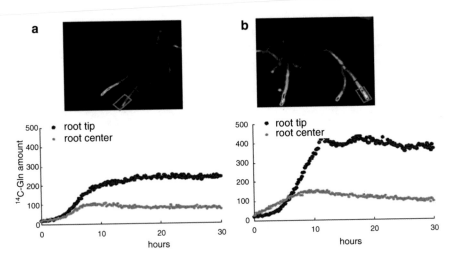

Fig. 4.20 Comparison of the ^{14}C-glutamine absorption manner of the root when Vaseline was pasted on the leaves [9]. ROIs (region of interest) were set at the root tip and center of the root to plot the ^{14}C-glutamine absorption speed. When Vaseline was applied to the leaves (**a**), the absorption amount and speed of ^{14}C-glutamine were lower than those of the control (**b**). The increase in ^{14}C-glutamine ceased at approximately 12 h, which suggests that most of the ^{14}C-glutamine supplied to the solution has been absorbed by the root by this time

accumulation amount of glutamine in roots decreased (Fig. 4.20), suggesting that the accumulation amount was dependent on transpiration activity. A similar result was obtained when valine or alanine was supplied to the rice plant from the root, but the amount of absorbed glutamine was much higher than those of valine or alanine (data not shown).

Further tracer work using doubly labeled (^{15}N and ^{13}C) glutamine and valine showed the difference in the availability of organic nutrients in a rice plant. When the absorption of glutamine and valine was compared, in the case of glutamine, the accumulation of ^{13}C in the plant was lower than the absorbed amount although the absorption and accumulation amounts of ^{15}N in the plant were the same. It was suggested that a portion of the amino acids assimilated into the plant was lost as carbon dioxide ([13]CO_2) through respiration, and the assimilation of glutamine occurred more smoothly than that of valine.

4.2.4   Introduction of the Plant Irradiating System (Second Generation)

4.2.4.1   Introduction of a Plant Box

Since light irradiation on the aboveground part of the plant is needed during imaging, a plant box was prepared using aluminum plates and set in the prototype imaging

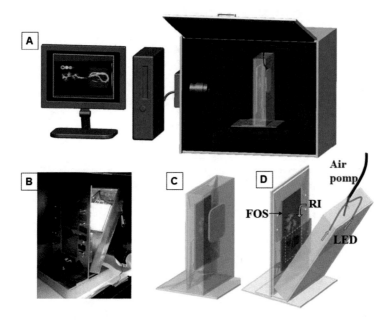

Fig. 4.21 Preparation of a plant box for RRIS imaging [10]. A plant box (200 × 300 × 50 mm) was prepared with a 5-mm-thick Al plate with a rectangle window of 100 × 100 mm. One hundred LEDs were installed in the box to irradiate the plant. All plates were sealed well with Al tape to prevent complete light leakage

box [6, 10–12]. The container was a rectangular box (50 mm × 200 mm × 300 mm). Figure 4.21 shows a picture of the plant container equipped with 100 light-emitting diodes (LEDs) capable of emitting 120 μmol/m^2 s onto the sample plant. By completely covering the plant and the light source, the container prevented external light from producing noise during the imaging processes. Since the scintillator allows light penetration, its surface facing the container was covered with an aluminum sheet 50 μm in thickness. Under this condition, light emitted by the LEDs was completely shielded, and only the β-rays penetrating the aluminum plate reached the scintillator without noticeable deterioration. The size of the scintillator working surface on the FOP was 10 cm × 20 cm. The LEDs could be switched on and off. The conditions of day and night were prepared by turning the LEDs on and off to produce a 16 h L/8 h D light/dark cycle for the aboveground part of the plant, while the roots were kept continuously in the dark. To control the temperature and humidity in the box, airflow was introduced into the box from the upper part.

The first experiment using the prepared plant box examined ^{32}P-phosphate absorption in *Lotus japonicas* cv. Miyakogusa. Phosphate absorption images were taken after the application of ^{32}P-phosphate (1 MBq/30 mL Hoagland culture solution) to the plant, and images were accumulated every 3 min for 30 h. The successive absorption images of phosphate from root to the aboveground part are

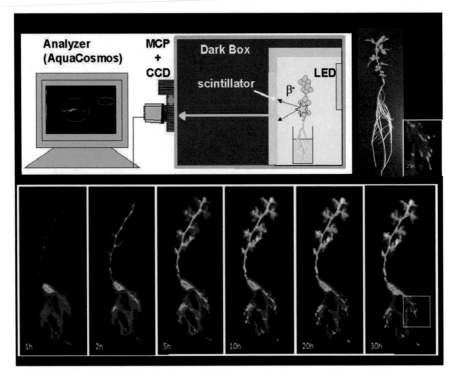

Fig. 4.22 Successive ^{32}P-phosphae absorption images in lotus [6]. The plant after 25 days of germination was set in a plant box, as illustrated in Fig. 21. ^{32}P-phosphate solution was supplied, and absorption images were taken 30 h after the treatment. The accumulation time for each image was 3 min. The magnification image of the root tips after 25 h is shown in the *upper right*

shown in Fig. 4.22. After 30 min of the ^{32}P-phosphate supply, the ^{32}P signal had reached the top of the stem. In the case of the root, high accumulation of phosphate in root tips was always shown throughout the successive absorption images, suggesting that a high amount of phosphate was always required in actively proliferating tissue, similar to that of the N source shown in Fig. 4.19. A magnification of the root tip is also shown in the figure.

To determine the difference in phosphate accumulation among the tissues in the plant under different light conditions while the plant was grown under 16 h L/8 h D light/dark conditions, the absorption amount of ^{32}P-phosphate in different tissues during development was plotted. In the flowers, the absorption curve was indifferent to the light cycle, but in the leaves, the absorption curve became high in light and low in the dark, where the opposite tendency of the absorption curve was observed in the roots (Fig. 4.23).

An interesting thing to note regarding phosphate absorption was that the amount of phosphate transferred to younger tissue was always high; however, under phosphate-deficient conditions, the amount of phosphate transferred to elderly leaves was similar to that transferred to younger leaves. Another interesting finding was that

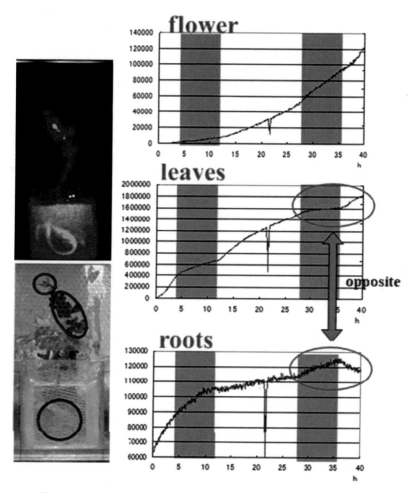

Fig. 4.23 ^{32}P-phosphate uptake from lotus roots. The plant sample was set in a plant box and grown under 16 h L/8 h D light/dark conditions. During 40 h, successive ^{32}P-phosphate uptake images were taken, and the ^{32}P signal amounts in 3 different tissues, shown as *red circles,* were removed and plotted. *Purple* columns in the figure show *dark periods*. The absorption *curve* shows that the flower parts were indifferent to the light cycle, and leaves absorbed high amounts of ^{32}P-phosphate during the light period, which was opposite to the behavior of the roots. The ^{32}P uptake was indifferent to the light cycle in flowers and increased during the light period in leaves but increased during the dark period in roots. The *vertical axis* shows the intensity of the counts

under phosphate-deficient conditions, morphological differences in root shape were induced, which decreased the amount of phosphate in the roots although there was no difference in root weight. The images obtained by the RRIS showed that it is a promising tool to trace ion transport, providing us with many new questions to be studied. This study was further developed to identify the kind and role of phosphate transporter genes under different conditions and in different kinds of plants. In particular, this study was further developed to study the expression of 7 phosphate transporter genes, from *LjPT1* to *LjPT7*, and the expression of each transporter gene in different tissues at different developmental stages was investigated (data not shown). However, the details are omitted here.

4.2.5   Introduction of Dark Period while Acquiring the Image (Third Generation)

The first generation of the RRIS system was applicable only under light-free conditions to protect the highly sensitive charge-coupled devised camera; therefore, the plant was not able to maintain suitable photosynthesis activities. The second generation of the RRIS system enabled experiments under light irradiation by incorporating a plant box. This system was able to detect both high-energy beta emitters (e.g., ^{32}P) and X-ray/γ-ray emitters (e.g., ^{109}Cd). However, since the irradiated light was able to penetrate the scintillator and this light increased the background noise of the photon-counting camera, it was necessary to install an aluminum shield (50 µm in thickness) over the scintillator. Since β-rays from low-energy β-ray emitters (e.g., ^{14}C, ^{35}S, ^{45}Ca) were not able to pass through the aluminum shield, these radioisotopes were not detected by this system. Among the low-energy β-ray emitters, the detection of ^{14}C under light-irradiated conditions was required to study the movement of photosynthetic products. Other nuclides, S and Ca, are major essential elements for plants; therefore, ^{35}S and ^{45}Ca are also important radioisotopes to analyze. Consequently, the system was improved to detect low-energy β-ray emitters.

Figure 4.24 shows a comparison of the second RRIS and new third generation RRIS. Instead of shielding the photon-counting camera from the light-emitting diode (LED) light with an aluminum shield and inner light-tight box, the new system was able to turn off the LED lights during photon counting. The detailed explanation is that the lighting system was integrated with the photon-counting camera control system. The power to the LED light was shut off by using a PC-controllable relay that was controlled by the AQUACOSMOS photon-counting camera control system via RS-232C. With this integrated control system, even if the system was accidentally stopped at any step of the sequence, the photon-counting camera was not damaged by LED light. The clustered LED light was composed of 64 red and 6 blue LEDs. These 2 types of LEDs had suitable peak wavelengths for photosynthesis (660 and 470 nm). Plants are more sensitive to the beginning of night than to daybreak; therefore, the properties of light, such as wavelength, should be carefully considered. Another device was the ventilation of the light-tight box. Opaque tubes were connected to the light box, and the air in the box was circulated at the laboratory temperature, which was maintained in a suitable range for the plants.

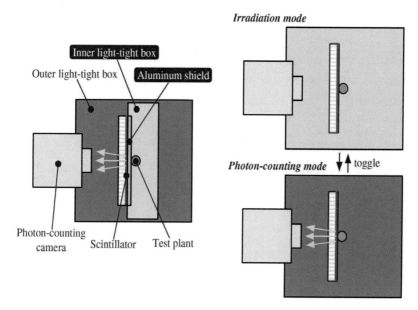

Fig. 4.24 Comparison of the previous RRIS (second generation) and new RRIS (third generation) [13]

With this change, both the detection of low-energy β emitters and light irradiation of the plant could be performed. In addition, a commercially available digital camera was added to the system to acquire photographic images of the test plant (Figs. 4.25, 4.26). The camera was placed opposite the photon-counting camera and was operated by remote control. This change enabled the continuous comparison of photon-counting images and photographic images throughout the experimental period.

^{35}S-labeled sulfate was chosen as the radioisotope tracer to verify the new system's ability to detect weak β-rays. For the test experiment, a rice plant (*Oryza sativa* L. var. Nipponbare) was grown for 18 days, and then carrier-free ^{35}S was supplied to 30 mL of culture medium, which contained approximately 1 mM sulfate. The specific radioactivity of ^{35}S was 170 kBq/ μmol.

The cycle of intermittent lighting was set to 1 h, and the photon-counting time in each cycle was set to 15 min. Figure 4.27 shows the image of the ^{35}S absorbed in the third, fourth, and fifth leaves of the two rice plants. The intermittent lightning system maintained the plant in healthy condition for several days. The increasing intensity of ^{35}S in the images showed that the test plants continued to absorb sulfate for 72 h after the application. As shown in Fig. 4.27, the third leaf absorbed sulfate quickly, but the distribution of ^{35}S was limited to the base area of the leaf blade. In contrast, in the fourth leaf, ^{35}S tended to be distributed over the entire area of the leaf blade. By superposing the radiation image on the picture, it was found which part of the leaf accumulated ^{35}S. In the fifth leaf, ^{35}S was detected later than in the other leaves, but the final intensity of the signal was higher than in the others. However, the increase

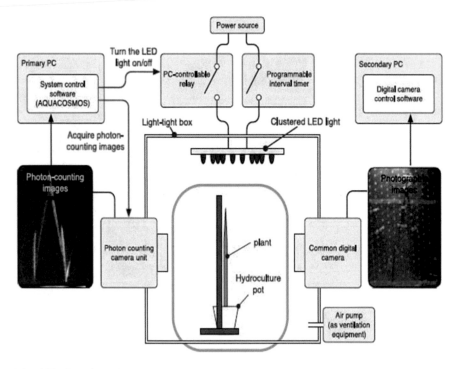

Fig. 4.25 Overview of the new RRIS [13]

in the signal area in the fifth leaf does not show the specific accumulation site of ^{35}S within the leaf; the fifth leaf itself grew larger during the imaging, which could not be determined from the RRIS image alone.

A revised RRIS with the ability to trace low-energy β emitters, such as ^{14}C, ^{35}S, and ^{45}Ca, was developed, and the capability of the new RRIS was verified. In particular, an advantage of the new RRIS is the ability to superimpose time-course photon-counting images on the photographic images of plants simultaneously. With this third-generation RRIS, another goal emerged: to image ^{14}C-labeled carbon dioxide gas and ^{14}C-labeled metabolites for the practical study of photosynthesis. (See next chapter).

4.2.6   Large-Scale Plant Sample

4.2.6.1   Plastic Scintillator

In all the generations of RRIS developed, a Cs(Tl)I scintillator deposited to a fiber optic plate (FOS) was used to convert radiation into light. However, one unit size of the scintillator was fixed at 10 cm × 10 cm, which was too small to observe an entire

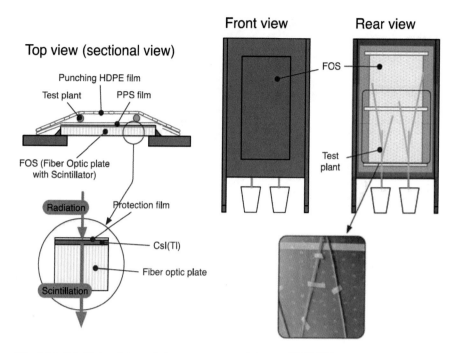

Fig. 4.26 Scintillator plates and plant arrangement of the new RRIS [13]

plant of larger size. To cover a large area of the sample, several FOSs were connected to each other. However, the plant samples sometimes grow much larger than the scintillator. For example, a rice plant can be as high as 50–60 cm, and it is difficult to prepare many expensive FOSs to cover the area of the whole plant.

To image large plants, six types of low-priced plastic scintillators were investigated: FOS, BC-400, BC-408 (Saint-Gobain, La Défense Cedex, France), and Lumineard-A, B, C, and D (Tokyo Printing Ink Mfg. Co., Ltd., Tokyo Japan). The thicknesses of the scintillators were as follows: FOS: 0.1 mm, BC-400: 0.5 mm, BC-408: 5 mm, Lumineard-A: 0.5 mm, Lumineard-B: 1 mm, Lumineard-C: 1.3 mm, and Lumineard-D: 5 mm. Lumineard-B and Lumineard-D were made from 2 and 10 sheets of Lumineard-A glued together, respectively. As an optical adhesive, a mixture of KE-103 and CAT-103 (Shin-Etsu Chemical Co., Ltd. Tokyo, Japan) at a ratio of 1:20 was used.

Then, the performance of the plastic scintillators was studied by preparing the standard solution. Two microliters of the ^{14}C-labeled sucrose solution (6–51,000 Bq/ spot) were spotted on a polyethylene terephthalate (PET) sheet. The size of the spots was approximately 3.6 mm^2, and the performance of these plastic scintillators, such as quantification, detection limit, and resolution, was studied and compared to that of FOS. Table 4.5 shows the properties of each scintillator.

The lower quantification limits in all scintillators (Bq/mm^2) were approximately constant, regardless of the integration time. Among the 6 plastic scintillators, the

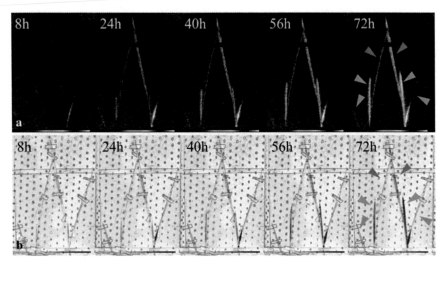

▶ 3rd blade   ⬈ 4th blade   ⬈ 5th blade

Fig. 4.27 Sequences of successive images of test plants acquired using the new RRIS [13]. ^{35}S-sulfate (170 kBq/μmol) was supplied to 30 mL of culture solution, and successive accumulation images of ^{35}S in two rice plants were taken for 72 h after the treatment. The upper sequence a shows ^{35}S images of ^{35}S-sulfate absorbed by the rice plant, and the pseudocolor indicates the signal intensity (*red* represents high intensity). The lower sequence b shows the superimposed image. *Blue* gradational images represent the ^{35}S image, and *grayscale* images represent photographic images that were processed using Sobel filter. By superposing the images, we found that the increase in ^{35}S in the fifth leaf was not due to a change in accumulation pattern but due to the growth of the leaf, which was not known only from ^{35}S images

Lumineard-C showed the lowest quantification limit. Since plastic scintillators are not completely transparent, optical refraction seemed to occur in the BC-400 and BC-408. The interaction between the Lumineard-A and the low-energy β-rays irradiated from ^{14}C seemed weak because the Lumineard-A had the lowest thickness of 0.5 mm. To improve the interaction efficiency, the Lumineard-B and Lumineard-D were prepared by folding several sheets of Lumineard. However, the light converted by the 5 mm-thick Lumineard-D spread while passing through the scintillator, resulting in a decrease in the signal intensity. Considering all the performances, the Lumineard-C was selected for imaging ^{14}C in a large sample. The ^{14}C images acquired by the plastic scintillator are presented in the next section, where ^{14}CO$_2$ gas fixation images of a rice and a corn plant are shown.

In the case of ^{32}P imaging with a plastic scintillator, similar evaluation tests were performed using a ^{32}P spot (12.2 Bq/cm^2). The resolution was obtained from the line profile of the spot. The FWHM became wider in the order A, B, C among the scintillators. Considering other results, the study demonstrated that the 1 mm thick

Table 4.5   Features of the plastic scintillators and FOS [14]

	Integration time (min)	The lower limit (Bq/mm^2)	The upper limit (Bq/mm^2)	R-squared	Light output relative to FOS (%)
FOS	3	1×10^0	2×10^1	0.9965	–
	5	1×10^0	2×10^1	0.9971	–
	10	6×10^{-1}	2×10^1	0.9980	–
	15	1×10^0	7×10^1	0.9991	–
BC-400	3	2×10^0	6×10^2	0.9944	38
	5	1×10^0	6×10^2	0.9947	37
	10	1×10^0	6×10^2	0.9943	35
	15	2×10^0	6×10^2	0.9947	33
BC-408	3	2×10^0	6×10^2	0.9915	31
	5	1×10^0	6×10^2	0.9925	31
	10	1×10^0	6×10^2	0.9937	30
	15	2×10^0	6×10^2	0.9938	27
Lumineard-A	3	1×10^0	6×10^4	0.9924	50
	5	1×10^0	6×10^2	0.9922	49
	10	1×10^0	6×10^2	0.9926	47
	15	2×10^0	6×10^2	0.9963	45
Lumineard-B	3	1×10^0	3×10^2	0.9987	66
	5	1×10^0	3×10^2	0.9988	65
	10	1×10^0	3×10^2	0.9987	63
	15	1×10^0	3×10^2	0.9980	57
Lumineard-C	3	3×10^{-1}	3×10^2	0.9933	56
	5	3×10^{-1}	3×10^2	0.9920	56
	10	3×10^{-1}	3×10^2	0.9919	52
	15	3×10^{-1}	3×10^2	0.9950	49
Lumineard-D	3	2×10^0	6×10^2	0.9993	33
	5	2×10^0	6×10^2	0.9991	32
	10	2×10^0	6×10^2	0.9991	31
	15	5×10^0	6×10^2	0.9990	28

Lumineard-B has the best performance among low-cost plastic scintillators and can visualize and quantify ^{32}P in the RRIS despite lower light output and resolution.

4.2.6.2   Images Obtained by a Plastic Scintillator

Using the Lumineard-B as a plastic scintillator for ^{32}P imaging, the uptake of ^{32}P-phosphate in a rice plant was visualized. A 27-day-old rice plant (*Oryza sativa*

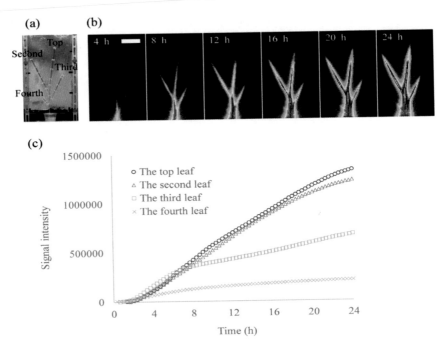

Fig. 4.28 ^{32}P-phosphate absorption in a rice plant using a plastic scintillator, Lumineard-B [14]. (**a**) Photograph of the rice plant set on Lumineard-B. (**b**) Sequential ^{32}P images taken by RRIS. *Scale bar*: 10 cm. (**c**) Time course of ^{32}P signal intensity in each leaf. The leaves were chronologically defined as the top leaf, second leaf, third leaf, and fourth leaf beginning from the youngest leaf, as indicated in the photograph

L. Dongjin) was fixed on a Lumineard-B converter covered with 2 μm thick aluminum film. The size of the Lumineard-B was approximately 800 mm × 200 mm. After the plant was transferred to 40 mL of culture medium containing 15 MBq of ^{32}P-phosphate, continuous RRIS imaging was performed in the dark box for 24 h. Light was provided at intervals of 10 min, and images were taken during the dark period. The time-course movement of ^{32}P-phosphate in a rice plant was measured by setting ROIs at each leaf (Fig. 4.28). The leaves were assigned as the top leaf, second leaf, third leaf, and fourth leaf in chronological order from the youngest leaf. There was a drastic change in the signal intensity according to the age of the leaf, where the ^{32}P signal intensity was the highest in the youngest leaf, indicating a high requirement of phosphorus for growth. Furthermore, ^{32}P signal intensity increased monotonically in each leaf. However, the rate of increase declined gradually, possibly because phosphorus in the leaves was translocated to other younger tissues. This decline was remarkable, particularly in the third leaf at 7 h after the treatment.

4.2.7   Summary of RRIS Development

A real-time RI imaging system (RRIS) was developed, composed of a Cs(Tl)I scintillator deposited on a fiber optic plate (FOP) and a highly sensitive charge-coupled (CCD) camera with an image intensifier unit. The imaging system was developed in 3 steps to enable irradiation of the plant during visualization. With the third generation of the system, it was possible to visualize the uptake behavior of many kinds of nuclides, such as ^{14}C, ^{22}Na, ^{38}Mg, ^{45}Ca, ^{32}P, ^{65}Zn, ^{86}Rb, ^{109}Cd, and ^{137}Cs, in a plant. It was shown that even among the leaves or within a root, the routes of element transfer and the accumulation behavior were different.

The development of the imaging system is summarized as follows (Fig. 4.29):

1. First generation: The scintillator selected was Cs(Tl)I, which was deposited on a multichannel plate. Everything, including the plant sample, was set in a dark container.
2. Second generation: A plant box was prepared so that only the aboveground part received light irradiation. The FOS had to be covered with Al foil to prevent light penetration, which prevented the counting of low-energy β-rays.
3. Third generation: Weak radiation energy could be detected, such as ^{14}C or ^{35}S. The light was off when the CCD camera was working. A camera was set to take a picture on which the radiation image could be superposes. $^{14}CO_2$ gas was generated and supplied to the plant. The fixed gas in the plant could be imaged.

Although RRIS has been used to study element movement in plants, its use has been limited to small plants because of its small field of view (100 × 200 mm). Therefore, the RRIS has been further updated to image an RI in a large plant. The study demonstrated that 1 mm thick Lumineard-B and Lumineard-C had the best performance among low-cost plastic scintillators and could visualize and quantify ^{32}P and ^{14}C, respectively, in RRIS despite lower light output and resolution than those of FOS. As a result, we are now equipped to analyze phosphorus movement in larger plants for advanced growth stages. This updated RRIS has a field of view of approximately 500 × 600 mm.

4.3   Element Absorption from Roots

4.3.1   Water Culture and Soil Culture

4.3.1.1 ^{32}P-Phosphate Absorption in a Rice Plant

Since β-rays from ^{32}P (1.7 MeV) can penetrate soil, it was possible to use the RRIS to visualize how phosphate in soil can be absorbed by roots. To compare the soil culture, water culture was also performed, and phosphate uptake images were taken.

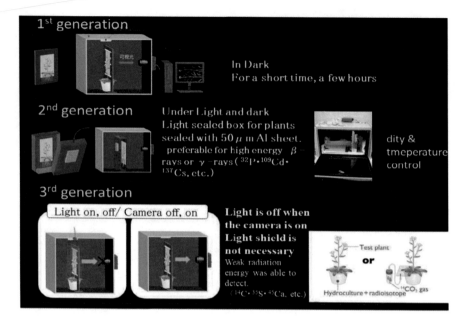

Fig. 4.29 Development of the real-time RI imaging system [7]. The sample for imaging and a scintillator device were kept in the dark in the first generation, since the CCD camera employed to image the light from the scintillator is highly sensitive to light. Then, the light-shielded plant box was prepared; therefore, light could irradiate the aboveground part of the plant in the second generation. In the third generation, the light was off when imaging was performed, so that weak radiation energy could be detected. The third generation enabled us to image the behavior of $^{14}CO_2$ gas and ^{14}C-photosynthate

Figure 4.30 shows the successive ^{32}P-phosphate uptake images of 3-day-old rice seedlings (*Oryza sativa* cv. Nipponbare) during 60 h of culture in soil and water. One of the rice plants was grown in 22.5 mL of water culture solution (Hoagland medium) containing 1.5 MBq of ^{32}P-phosphate. The other was grown in 20 mL (32 g) of nursery soil for rice seedlings (Kumiai Baido, Kasanen Industry Co, Japan, 3.1 g phosphate per 20 kg), which was mixed well with 15 mL of culture solution containing 1.5 MBq of ^{32}P-phosphate. The integration time for each imaging frame was 3 min.

In the water culture, the rice plant continuously absorbed higher amounts of ^{32}P-phosphate and grew much faster than the plant growing in soil. In contrast, in the soil culture, only a small amount of ^{32}P-phosphate was absorbed from the roots since phosphate was firmly adsorbed in the soil. In the soil culture, it was also observed that hardly any ^{32}P-phosphate was transferred to the aboveground parts, even after 20 h, and the growth of the plant was very slow. Because of the nature of the phosphate ion, which is weakly mobile owing to a very low coefficient of diffusion (10^{-12} to 10^{-15} m^2/s), the uptake of phosphate from the soil created a depleted area around the root. This depletion zone was observed as a dark colored

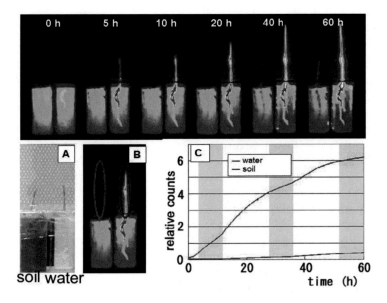

Fig. 4.30 Comparison of the ^{32}P-phosphate uptake by rice seedlings between water and soil culture [15]. Successive images of ^{32}P-phosphate uptake by the rice seedlings during 60 h of water and soil culture are shown. The integration time for each imaging frame was 3 min. For each record, the sample grown in soil is on the left, and the one grown in the water culture solution is on the right. (**a**) Picture of the sample. (**b**) Two ROIs (region of interest) in the ^{32}P image. The *blue* and *red* ROIs are the aboveground parts of the rice grown in soil and water, respectively. (**c**) ^{32}P-phosphate uptake curve in two ROIs. The *gray* columns in C are the dark period. Pseudocolor was assigned to the image according to the intensity of the radioactivity

area at the root, in the shape of the root, and clearly demonstrated that the phosphate adjacent to the root was taken up by the root. This depletion zone appeared within a few hours, and this area induced further movement of phosphate toward the root, as revealed by the increase in phosphate uptake.

Imaging by the RRIS also offers a way to investigate the effects of light and/or circadian rhythms on phosphate uptake. The light conditions of the plant were a 16 h L/8 h D light/dark cycle during the 60 h of imaging. When the amount of phosphate at the ROI in water culture, as indicated in Fig. 4.30, was plotted successively, the phosphate uptake clearly increased during the daytime. Similar observations have been performed with other plants, such as *Lotus japonicas* (data not shown). Indeed, light conditions or circadian rhythms have been found to directly or indirectly affect ion uptake, and such phenomena could have multiple and complex origins.

It is generally known that when the growth of plants in water culture and soil culture is compared, the growth of the plant grown in water culture is very fast compared to that of a plant grown in soil. This is one of the reasons water culture is employed in factories to grow vegetables; however, it is also known that cereals, including rice, have much higher yield when grown in soil; therefore, indoor factories are not suitable to grow rice or wheat.

4.3.1.2 ^{137}Cs Absorption in a Rice Plant

After the Fukushima nuclear accident, the movement of ^{137}Cs in soil attracted attention. Although we found that ^{137}Cs was adsorbed firmly on the clay in soil, the absorption of ^{137}Cs is sometimes studied using plants growing in water culture. Therefore, it was necessary to visualize the ^{137}Cs uptake behavior of rice plants growing in soil and to compare it to that of rice plants growing in water culture, partly to show the results to people who are concerned about the contamination of plants grown in contaminated soil.

The rice seedlings with three expanded leaves were grown in 3 mL of liquid medium or soil medium (3 g of soil plus 3 mL of liquid medium) containing 50 kBq of ^{137}Cs. The soil was collected from a paddy field in Fukushima district. The soil contained no radiocesium derived from fallout because the soil was collected from the deep part of the paddy field (5–10 cm from the surface). Therefore, in both cases, ^{137}Cs was supplied to both water and soil.

As shown in Fig. 4.31, the rice absorbed a high amount of ^{137}Cs from the root when cultured in water medium, whereas ^{137}Cs was hardly absorbed from the soil, which showed the same phenomenon as that of the ^{32}P supply cited in the previous section. When the radioactivity of liquid and soil was plotted, it was also confirmed that the radioactivity in soil hardly changed with time. The movement of ^{137}Cs uptake in both media was also presented as a movie.

In the case of water culture, the rate of increase in the ^{137}Cs content in leaves declined significantly in several hours, showing that the absorption of ^{137}Cs was completed. The first curve of rapid translocation of ^{137}Cs to the shoot within 5 h could be interpreted as the xylem loading activity, and the subsequent slowly increasing curve that appeared after 5 h may be explained by the ^{137}Cs remobilization activity.

This research was further developed to study ^{137}Cs uptake and xylem loading activities using plants grown under K-deficient conditions. The other findings were that the accumulation of ^{137}Cs within the rice plant during the developmental stage was different from that in a soybean plant, which resulted in a higher ^{137}Cs amount in the edible part of the soybean plant than that of the rice plant (data not shown).

4.3.2   Multielement Absorption

4.3.2.1   Multielement Absorption Images in Arabidopsis by RRIS

Since the RRIS system enabled visualization of the element absorption behavior, live imaging of the other nuclides is shown. The first presentation is the visualization

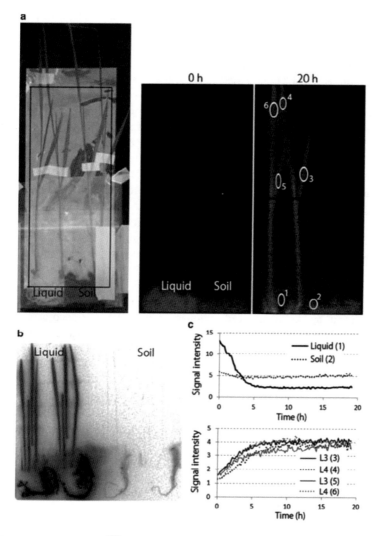

Fig. 4.31 Comparison of the ^{137}Cs uptake by the rice seedlings between water and in soil culture [16]. (**a**) Rice seedlings were set in the plant box, and real-time images were acquired by RRIS for 20 h. The integration time for each imaging frame was 10 min. The seedlings were grown in water culture until they developed three expanded leaves (L2, L3, and L4) under 16 h L/8 h D, light/dark conditions at 30 °C. Then, two seedlings were grown in 3 mL of liquid medium or soil medium (3 g of soil plus 3 mL of liquid medium) containing 50 kBq ^{137}Cs. The soil was non-contaminated and collected from a paddy field in Fukushima district. In the real-time images, six ROIs were set. ROIs 1 and 2 indicate the liquid medium component. ROIs 3 and 5 are the L3 blades, whereas ROIs 4 and 6 are the L4 blades. (**b**) IP images of the rice seedlings after the RRIS imaging were completed. The ^{137}Cs signal was hardly detected in the L2 and L3 sheaths. (**c**) ^{137}Cs accumulation in the six ROIs

of long-distance ion transport in Arabidopsis using radioisotope tracers, $^{22}Na^+$, $^{28}Mg^{2+}$, ^{32}P-phosphate, ^{35}S-sulfate, $^{42}K^+$, $^{45}Ca^{2+}$, $^{54}Mn^{2+}$, $^{65}Zn^{2+}$, $^{109}Cd^{2+}$, and $^{137}Cs^+$, supplied from the roots. Seeds of *Arabidopsis thaliana* Col-0 were grown in full-nutrient culture solution at 22 °C under 16 h L/8 h D light/dark conditions with 100 µmol/m^2 s of light. After 43 days, plants approximately 25 cm in height were selected and transferred to 20 mL of culture solution containing radioactive tracers of individual nutritional elements. The tracer concentrations applied were as follows: $^{22}Na^+$, 25 kBq/mL; $^{28}Mg^{2+}$, 25 kBq/mL; ^{32}P-phosphate, 50 kBq/mL; ^{35}S-sulfate, 500 kBq/mL; $^{42}K^+$, 1 kBq/mL; $^{45}Ca^{2+}$, 250 kBq/mL; $^{54}Mn^{2+}$, 50 kBq/mL; ^{65}Zn, 75 kBq/mL; $^{109}Cd^{2+}$, 50 kBq/mL; and $^{137}Cs^+$, 25 kBq/mL. $^{42}K^+$ was prepared from an ^{42}Ar–^{42}K generator by milking. ^{28}Mg was produced by the ^{27}Al (α, 3p) ^{28}Mg reaction and separated from the Al target. The imaging area was the aboveground parts between 3 and 22 cm from the root. Samples were irradiated under a light intensity of 100 µmol/m^2 s for 15 min at intervals of 15 min. During each 15-min interval, imaging was performed without light. The radiation converted to photons by the scintillator was harvested for 15 min with a highly sensitive CCD camera (C3077–70, Hamamatsu Photonics Co.).

During 24 h of the radioactive ion supply, the accumulation pattern and uptake speed of each element exhibited specific features. From Figs. 4.32, 4.33, and 4.34, each element uptake image with time is shown, and pseudocolor was added according to the intensity of the radioactivity. When these successive figures in each element were connected, a movie was produced to show the movement.

Sequential analysis showed three distribution patterns in the aboveground part of Arabidopsis. The first was a widespread distribution over time, as exhibited by ^{22}Na, ^{32}P, ^{35}S, ^{42}K, and ^{137}Cs. The second pattern shown by elements ^{28}Mg, ^{45}Ca, and ^{54}Mn was a higher accumulation in the basal part of the main internode. The third pattern was that accumulation was only found in the leaf tips or the bottom parts of the flower, as in the case of ^{65}Zn or ^{109}Cd. In the case of ^{45}Ca, ^{28}Mg, or ^{54}Mn, only a small amount of the ion reached the tip of the stem, even after 24 h, and the movement was very slow. In contrast, the heavy elements ^{65}Zn and ^{109}Cd moved very fast, and when they were transferred to the aboveground part, they suddenly moved to the leaves without accumulating in other tissues and accumulated at the leaf tips. Representative examples of the three element profile patterns after 24 h are shown in Fig. 4.35.

The difference in the absorption images, indicating the differences in distribution and speed of movement, seemed to be derived, at least in part, from the chemical forms of the elements; one group comprised monovalent cations or anions such as ^{22}Na, ^{42}K, and ^{137}Cs, whereas the other group was multivalent cations, such as ^{28}Mg, ^{45}Ca, and ^{54}Mn. Given the widespread distribution profile along the main stem from the lower to the upper parts, monovalent cations and anions appeared to move through the vascular tissue smoothly and quickly, whereas multivalent cations moved slowly. The low velocity of multivalent cation transport is possibly derived from the interaction between the ions and the negatively charged cell wall of xylem vessels.

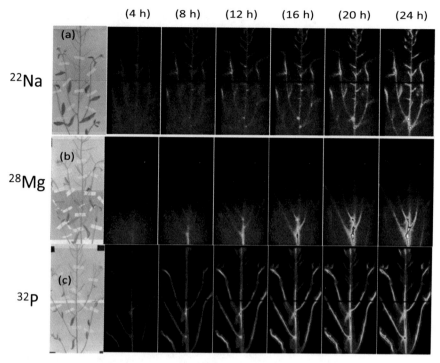

Fig. 4.32 Successive images of the ion movement in Arabidopsis taken by RRIS (1) [17]. ^{22}N (**a**), ^{28}Mg (**b**), and ^{32}P(**c**) were supplied to the roots. The detection time for each imaging was set to 15 min

4.3.2.2 Mg Movement in Arabidopsis

Among the elements investigated, the characteristic transport of ^{28}Mg within the main stem of the inflorescence is presented as an example. When two regions of interest (ROIs) were set at different parts of the internode, the difference in radioactivity counts with time shows the movement of the element more clearly. Two ROIs were set at an interval of 30 mm (Fig. 4.36). The signal intensity of ^{28}Mg in the ROI: A, which was set at the lower position, exceeded the limit of quantitation (LOQ) soon after the imaging was started and continued to increase linearly. The LOQ corresponds to the earliest time when the radioisotope was first able to be detected. Subsequently, after approximately 6 h, the ^{28}Mg content in ROI: B, which was set at the higher position, began to increase linearly. According to the time gap between ROI:A and ROI:B, the time required for ^{28}Mg to travel 30 mm was 5.5 h. Accordingly, the velocity of Mg^{2+} toward the top of the main stem was estimated to be 5.5 mm/h. By similar measurement, the velocity of ^{32}P was calculated to be >60 mm/h.

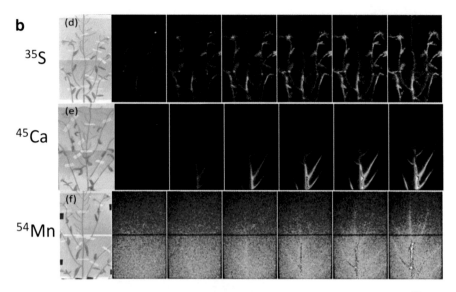

Fig. 4.33 Successive images of the ion movement in Arabidopsis taken by RRIS (2) [17]. ^{35}S (**d**), ^{45}Ca (**e**), and ^{54}Mn (**b**) were supplied to the roots

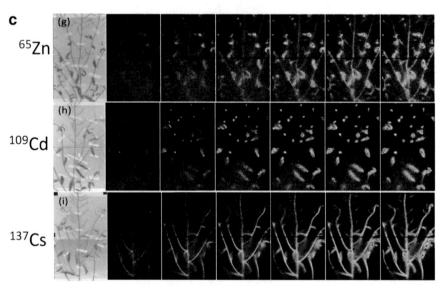

Fig. 4.34 Successive images of the ion movement in Arabidopsis taken by RRIS (3) [17]. ^{65}Zn (**h**); ^{109}Cd (**i**); ^{137}Cs (**j**)

The amount of ^{28}Mg delivered to the upper part of the plant was small. However, Mg is required at the tip of the main stem of the inflorescence. Together with limited information about the movement of Mg, the accumulation behavior at each node was

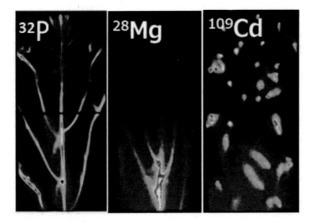

Fig. 4.35 Representative three patterns of the element profile after 24 h. ^{32}P group: widespread distribution throughout the plant; ^{28}Mg group: very slow movement, accumulated at the basal part of the main internode; ^{109}Cd group: very fast movement, accumulated at the leaf tips and bottom parts of the flower

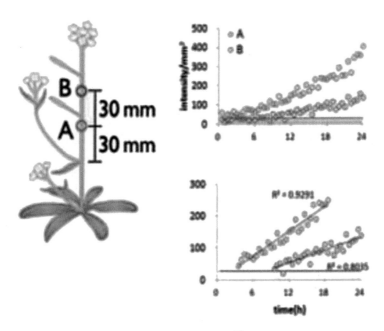

Fig. 4.36 (**d**) Time-course analysis of the radioactivity of ^{28}Mg detected in two ROIs [17]. ROI A: a *blue circle* was set on the main stem, 30 mm upper part from the top node. ROI B: a *red circle* was placed 30 mm above ROI A. The *solid line* shows the limit of quantitation (LOQ). The broken line shows the limit of detection. The linear components in the upper graph were extracted and are shown in the lower graph

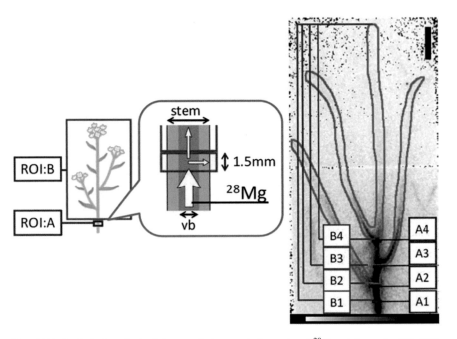

Fig. 4.37 Analytical method of the transfer/accumulation ratio of ^{28}Mg at the nodes [3]. ROIs A1–A4: each 1.5 mm in length, and the number increased with height. A1 was set at the internode, and A2–A4 were set at the nodes. ROIs B1-B4: each ROI B is the upper part of the corresponding ROI A area. *vb* vascular bundle

analyzed [3]. To estimate whether Mg accumulated at this part for transfer to the connected branch or was delivered only to the upper main stem, the ratio of Mg accumulated at a certain part of the node to all the Mg absorbed above this node was calculated. As shown in Fig. 4.37, one internode and three nodes above this part, each 1.5 cm in length, were selected (A1–A4), and the part of the plant higher than each position of A was selected (B1–B4). At all sites, A1 to A4 and B1 to B4, the absorption curve linearly increased (Fig. 4.38). Then, the accumulated ratio at each A position was calculated as A/(A + B). From this ratio, the loading manner of Mg from xylem tissue could be estimated, namely, when the ratio is high, the accumulation character as a sink is high and the low ratio indicates the low accumulation amount at the site and Mg was just transferred to the upper part of the plant. When the ratio was calculated, there was no change in the ratio of the accumulation at each node with time, and the accumulation rate increased as the position of the node became higher (Fig. 4.39). The results indicated that the ratio of Mg transfer from xylem tissue at the node was kept constant, and the accumulation rate was higher at the upper part of the stem, suggesting that the speed of the transfer movement of ^{28}Mg decreased as the height of the internode increased.

However, after the ions reach the bottom part of the shoot, the part played by the phloem in promoting ion transport should be considered in addition to xylem flow. To evaluate phloem contribution to ion transport, heat girdling was performed for

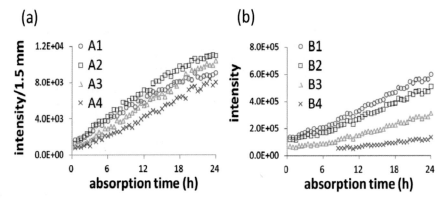

Fig. 4.38 ^{28}Mg intensity with time at ROIs A-B in Fig. 4.37 [3]. (**a**) ^{28}Mg intensity in ROIs A1-A4; (**b**) ^{28}Mg intensity in ROIs B1-A4

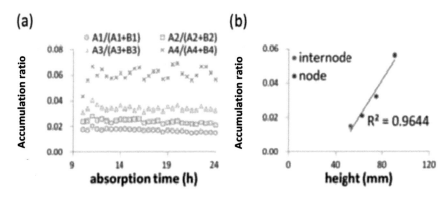

Fig. 4.39 Accumulation ratio of ^{28}Mg at ROIs An in Fig. 4.37 [3]. (**a**) accumulation ratio at the internode (A1) and nodes (A2–A4); (**b**) accumulation ratio of A1-A4 with height

Arabidopsis grown for the same period as used in the previous section, 43-day-old seedlings. The main stem was heated for several seconds by a soldering iron. First, 1 MBq of $^{14}CO_2$ was supplied to the rosette leaves, and the image of the whole plant was taken by IP to confirm the reliability of the heat-girdling technique. Since the IP showed no photosynthate image before or after the treatment in the internode of the plant (data not shown), 10 kBq/mL of ^{28}Mg was supplied to the plant, and the absorption curve was obtained from the successive images acquired by RRIS. The ^{28}Mg distribution pattern along the main stem was similar to that of nontreated plants (Fig. 4.40). A kinetic analysis showed that the velocity of Mg^{2+} in the xylem flow was 5.5 mm/h, a similar value to that found in intact Arabidopsis cited above. Thus, the upward Mg^{2+} movement within the third internode of the main stem is likely to be mediated mainly by xylem flow, while the phloem contribution is scarce during the first 24 h of root absorption. In contrast, in the case of ^{32}P-phosphate absorption,

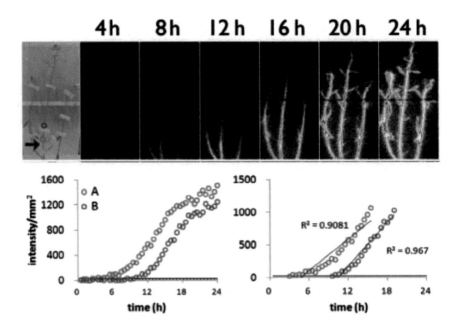

Fig. 4.40 ^{28}Mg uptake image of Arabidopsis after the heat-girdling treatment [17]. Heat-girdling treatment was performed at the position marked with a *red arrow* (*upper left*). The exposure time of the camera was set to 15 min. The ROIs are indicated by *blue circles* (ROIs A) and *red circles* (ROIs B) in the image. *Upper*: distribution images of ^{28}Mg within 24 h after the treatment. *Lower*: signal intensity of ^{28}Mg in ROI A and ROI B. The linear components were extracted (*lower right*)

heat girdling resulted in strong ^{32}P signal accumulation at the bottom of the main stem, which was never observed in untreated Arabidopsis (data not shown).

4.3.2.3   Mg and K Absorption in a Rice Plant

Rice plant seedlings after 12 to 14 days of germination were used to obtain uptake images of ^{28}Mg, ^{32}P, ^{35}S, ^{42}K, and ^{45}Ca for 12 h. ^{28}Mg and ^{42}K images are shown in Fig. 4.41.

The ROI (region of interest) was set at the third and fourth leaves, and the uptake of both ^{28}Mg and ^{42}K was plotted. It was shown that the accumulation patterns of both elements are different. The amount of Mg in both leaves was similar, and the amount increased linearly until 12 h. With further development of this study, applying this ^{28}Mg imaging technique, an early response of Mg deficiency was found to appear, especially in the fifth leaf [18]. This result was further examined to determine the mechanism of Mg absorption and translocation, using ^{28}Mg as a tracer (data not shown). In the case of K, there was a drastic difference in the accumulation amount between the third and fourth leaves. The absorption speed and the amount of K in the third leaf were more than two times higher than those in the fourth leaf,

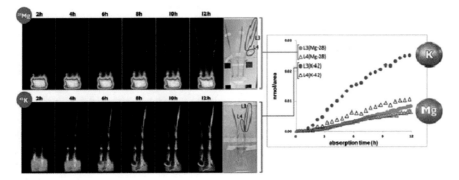

Fig. 4.41 ^{42}K and ^{28}Mg uptake images in rice plants [2] modified. The ROIs were set at L3 (third leaf) and L4 (fourth leaf). L4 grew after L3, and the signals at these ROIs were plotted

suggesting a quick K movement response, especially to tissue where the K requirement was high. Since K showed a competitive character for Cs absorption, application of K was found to inhibit ^{137}Cs absorption from the root when ^{42}K was applied as a tracer. ^{42}K was further used to study the mechanism of ^{137}Cs contamination caused by the Fukushima nuclear accident.

4.3.3 Summary of Element Absorption from Roots

Since radiation can penetrate both water and soil, it was revealed that there was a great difference between water culture and soil culture in growth as well as ion absorption from roots. It is rather popular to perform physiological studies of plants employing water culture; however, extending the results to understand plants growing in the field reveals discrepancies in physiology between plants grown in water and those grown in soil, i.e., the plant physiology is totally different.

For example, as cited above, it is generally known that the grain yield of rice grown in soil is much higher than that of rice grown in water culture. Although it is not known what determines the yield of the plant, the fact that slowly growing plants produce higher amounts of seeds might suggest that the different usage of energy by the plant during growth, such as the higher energy requirement for the absorption of phosphate from soil, might affect the production of many grains. The roots actually use large amounts of energy to remove phosphate from the closest soil to absorb. As a result, there was always ^{32}P-depleted zone whose shape reflected that of the root itself. Soil culture is very complicated because various physiological and biological factors are in the soil itself; however, for soil culture, the application of RI is an indispensable tool to study the physiological aspects of plants.

The real-time RI imaging system (RRIS) was applied to visualize multielement absorption in a plant. The absorption velocity is very different among the elements, which results in different distribution patterns within a plant. In the case of

Arabidopsis, the element-specific absorption patterns were clearly classified into three patterns within 24 h of root absorption. It was very interesting to note that so many ions with such different velocities are actually moving in water, which is also flowing upward from the root in the xylem.

Since the image was produced based on radiation, image analysis could be performed, and the case of ^{28}Mg is presented as an example. Through image analysis, it was possible to differentiate xylem flow from phloem flow. In the case of ^{28}Mg, most of the flow was through the xylem, and phloem flow was hardly observable within 24 h of transport. Applying this imaging method, the specific element accumulation sites were also detected.

4.4   Development of a Microscopic Real-Time RI Imaging System (RRIS)

The RRIS presented above was for a relatively large-scale sample, for example, a whole plant or whole tissue. However, to perform microscopic RI imaging, different types of solutions must be developed to enable the desired level of magnification. Therefore, a new system microscope-modified system was developed, integrating a thin Cs(Tl)I scintillation system and a magnification device.

4.4.1   Modification of a Fluorescence Microscope

The first step was the preparation of a scintillator. The scintillator thickness determines the resolution and sensitivity of the radiation image, where a thinner scintillator provides higher resolution; however, higher sensitivity is acquired with increasing thickness. The penetration of radiation through the scintillator is another factor to take into account, especially when the β-ray energy is high. Therefore, to install the scintillator for the microscope, the thickness of the scintillator (Cs(Tl)I) should be properly prepared with respect to the sensitivity and resolution. Several kinds of Cs(Tl)I scintillators were prepared with different thicknesses of Cs(Tl)I, 10, 25, 50, 100, and 200 μm, deposited on a fiber optic plate (FOS) in vacuum. The standard samples were 0.37 kBq of ^{32}P, 1.85 and 3.7 kBq of ^{45}Ca and 0.925 and 1.85 kBq of ^{14}C, prepared from $H_3{}^{32}PO_4$, $^{45}CaCl_2$, and ^{14}C-glucose solutions, respectively, and mounted on a membrane filter. The filter mounted with the spots was covered with a Mylar film (4 μm). Then, an FOS was placed on the filter with the standards, and the measurement was performed for 3 min by a GaAsP imaging intensifier unit with a detection area of 5 × 5 cm (See 4.2.2 of this chapter).

To address resolution, a membrane filter soaked with ^{32}P solution (3.7 kBq/ μL) was prepared. On this filter, two iron plates 1 mm in thickness were placed parallel to each other at a distance of 500 μm. Then, FOSs with different scintillator thicknesses

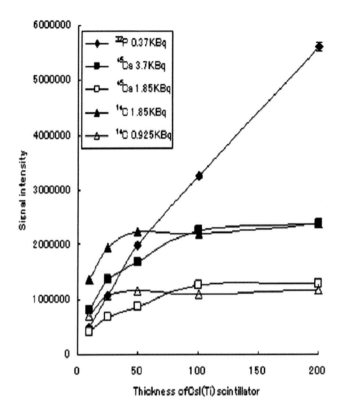

Fig. 4.42 Thickness of the Cs(Tl)I scintillator of the fiber optic plate (FOS) and detection efficiency [19]. Cs(Tl)I scintillators with different thicknesses were prepared by depositing on FOS in a vacuum. The standard solutions of ^{32}P, ^{45}Ca, and ^{14}C were prepared as spots on a membrane filter

were placed on the sample, and the β-ray image of the slit was analyzed. The detection limit was set to twice as high as the background intensity.

Figure 4.42 shows the relation between the thickness of the scintillator, Cs(Tl)I, and the detection efficiency. As shown in the figure, the detection efficiency depends on the energy of the radiation from the nuclide, where ^{32}P (β-ray energy max: 1.709 MeV) continuously increased with increasing thickness from 10 to 200 μm. In the case of ^{14}C (β-ray max: 0.156 MeV) and ^{45}Ca (β-ray max: 0.257 MeV), the detection efficiency plateaued after 50 and 100 μm, respectively. The resolution of the image increased with decreasing scintillator thickness (Fig. 4.43). A 50 μm thickness of the scintillator deposition was selected as preferable for imaging a nuclide with lower β-ray energy, such as ^{45}Ca or ^{14}C.

Once the thickness of the scintillator was decided, the next step was the magnification method. In the microscopic imaging system, a taper FOS was applied (Fig. 4.44), which allowed five times magnification. The diameter of the FOS was 3 μm on one side and 15 μm on the other side. The scintillator was deposited on the

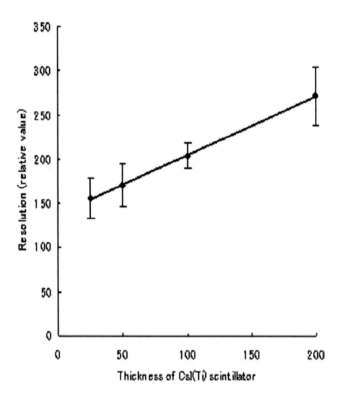

Fig. 4.43 Thickness of the Cs(Tl)I scintillator and image resolution [19]. Two parallel iron plates that formed 500-μm slits were placed on the RI sheet, and FOSs with different scintillator thicknesses were placed on the slit. The line profile of the slit image produced by the FOS was analyzed to obtain the resolution

surface of the smaller end. The surface of the tapered FOS was covered with an aluminum Mylar to prevent contamination as well as background light. As shown in the figure, the system consisted of the FOS, a lens, and a GaAsP imaging intensifier. To obtain higher-resolution images, an Axio Cam HRm (Carl Zeiss, Co.) was applied to acquire fluorescent images in the imaging intensifier unit, and Axio Vision (Carl Zeiss, Co.) was used for image analysis.

The transmitting light unit was removed, and a new pole was installed for the microscope so that the real-time imaging system could be smoothly shifted vertically by an electric motor. After bright field and fluorescence images were taken, the scintillator side of the tapered FOS was placed on the sample, and the radiation image was taken from 3 to 10 min of integration.

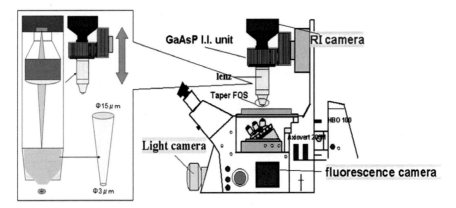

Fig. 4.44 Schematic illustration of micro-RRIS [19]. A fluorescence microscope was revised to acquire radiation images and fluorescence images. Taper FOS was prepared, which enabled five times magnification of the image, and the scintillator was deposited on the smaller end area. To acquire radiation images, a metal tube consisting of the taper FOS, lens, and GaAsP imaging intensifier was installed

4.4.2 Radiation Images Under the Modified Fluorescence Microscope

With the modification of the transmitting light unit, the fluorescence microscope could acquire three kinds of images: light image (BF camera), fluorescence image (FL camera), and radiation image (RI camera). Figure 4.45 shows a picture of the modified fluorescence microscope and an example of the three types of images acquired from the same soybean stem dissection sample when ^{45}Ca was supplied. The resolution of the ^{45}Ca image was not very high, since the thickness of the sample was 70 μm, which was the estimated size of a single cell. With this thickness, the β-ray irradiated from the sample was somewhat expanded before reaching the scintillator, resulting in an image with low resolution.

Under this modified fluorescence microscope, the distribution of different radio-isotopes (^{45}Ca, ^{35}S, and ^{55}Fe) in various tissues of *Arabidopsis* was observed, and the radioisotope images with pseudocolors were superimposed on the corresponding light images (Fig. 4.46). With the ^{45}CaSO$_4$ solution, images were obtained 1 h after supplying the solution to roots. A higher accumulation of ^{45}Ca was observed in younger leaves. On the other hand, in the case of ^{35}S, it took hours before ^{35}S was detected in leaves. The image of ^{35}S distribution in Fig. 4.46 was obtained after 48 h of Na$_2$^{35}SO$_4$ supply, and ^{35}S was observed to accumulate along the leaf veins.

As another example, Fig. 4.46 shows successive images of ^{55}F distribution acquired every 20 min after ^{55}FeCl$_2$ solution (100 KBq/25 μL) was supplied to a 3-day-old Arabidopsis seedling. The integration time of each imaging frame was 2 m. The figure illustrates the accumulation of ^{55}F in the root tip.

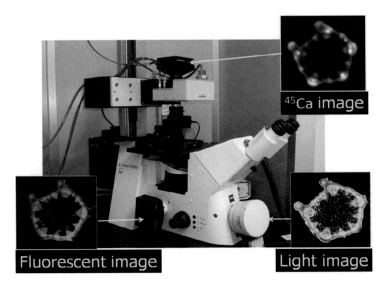

Fig. 4.45 Micro-RRIS with 3 types of images, radiation image, fluorescent image, and light image. ⁴⁵Ca solution (10 M Bq/20 mL, Ca: 20 μM) was supplied to the stalk of the first trifoliate leaves of a soybean plant after 3 weeks of germination. Then, the stalk was sliced to 70 μM, and a radiation image of ⁴⁵Ca was obtained. The neighboring slice was stained with fluorescent dye Fluo-3 M to acquire a fluorescent image of Ca

4.4.3   Further Modification of Micro-RRIS

Then, further improvement of the micro-imaging system was performed. To achieve a higher magnification, an optical lens was introduced instead of a tapered FOS. A combination of the FOS and an optical lens magnified the image 20 or 40 times, depending on the lens used (Fig. 4.47).

Since the image is based on radiation, quantitative analysis can be performed. To confirm the quantitative character of the image, the radioactivities calculated from the images of standard solution spots of ^{32}P and ^{35}S were compared with the actual radioactivities of the standard solutions prepared, ^{14}C, ^{55}Fe, ^{32}P, ^{35}S, and ^{109}Cd. In all cases, the linearity of the signal was conserved between the counts of the image (cpm) and the radioactivity of the mounted standard solution, which facilitated imaging over a broad range of concentrations. Out of 5 nuclides, Fig. 4.48 provides the results of ^{32}P and ^{35}S. As shown in the figure, in both cases, the linearity of the signal was conserved between the counts of the image (cpm) and the radioactivity of the standard solution prepared (Bq). Conservation of the linearity ranged from 0 to 6.5 kBq/ μL (^{32}P) or even to 28 kB/ μL (^{35}S). In all cases, such signal dynamics far exceed the signals used in experiments. It was shown that this micro-imaging system was thus applicable to a broad range of experiments. For example, in the case of ^{32}P solution, it was possible to detect the image of a 1 μL spot containing only 16 Bq in 2 min.

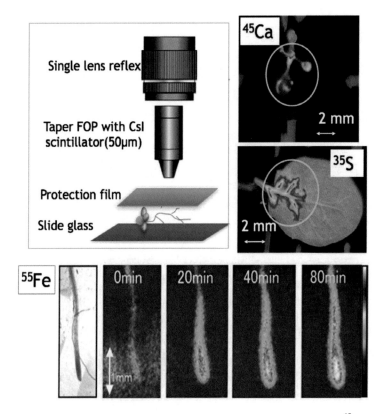

Fig. 4.46 Example of the images acquired by micro-RRIS in Arabidopsis [15]. For ^{45}Ca imaging, ^{45}CaSO$_4$ solution (1 MBq/0.5 mL) was supplied to the root of a 13-day-old seedling for 1 h. The Na$_2$^{35}SO4 solution (5 MBq/0.5 mL) was supplied to the root for 48 h. In both plants, imaging was performed after 10 h. To obtain successive images of the iron uptake, ^{55}FeCl$_2$ solution (100 kBq/ 25 mL) was supplied to a 3-day-old seedling; ^{55}Fe accumulation was observed in the root tip

To confirm the linear relation between the radioactive counting in the image of the plant and the actual radioactivity of the tissue measured by liquid scintillator, an Arabidopsis plant was employed. The plant was grown for 10 days in 1/10 MS medium without phosphate, after which ^{33}P-phosphate solution (6 kBq/µL) was supplied for 5 min, and the root was imaged for 5 min. After imaging, the root tip (2 mm) was digested, and the radioactivity of ^{33}P was measured by a liquid scintillation counter. Then, the amount of ^{33}P measured in the image of the root grown under phosphate-deficient conditions was compared to that of a root grown under sufficient conditions. When ^{33}P-phosphate was supplied to the culture solution, the amount of ^{33}P accumulated in the root tip grown under phosphate-deficient conditions was 7–8 times higher than that grown under normal conditions, which

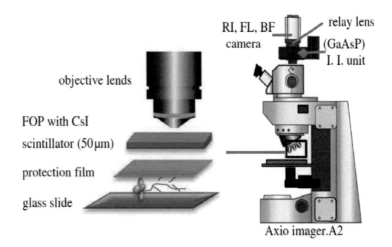

RI, FL, BF   relay lens
camera   (GaAsP)
              I. I. unit

objective lends

FOP with CsI
scintillator (50 μm)

protection film

glass slide

Axio imager.A2

Fig. 4.47 Modification of micro-RRIS [15]. To achieve a higher magnification, a combination of an FOS without taper and an optical lens was prepared, which magnified the image 20 or 40 times depending on the lens instead of using a taper FOS

was in good accordance with the actual data on harvested roots measured by a counter (Fig. 4.48).

Then, what kind of magnified root image could be shown under this further modified microscope? Figure 4.49 shows an example of a ^{32}P image of a 10-day-old seeding of Arabidopsis root when 100 Bq/μL of ^{32}P-phosphate was supplied for 5 min. The plant was placed on a glass slide, and a radiation image was taken. As shown earlier, phosphate was shown to accumulate in the root tip. When the root tip marked in the square was magnified and the light image was superposed with the ^{32}P image, the phosphate distribution in the root tip was visualized. In the past, an imaging plate (IP) was used to analyze phosphate uptake locations, and the root tips were identified as an important area for phosphate uptake. Nevertheless, the resolutions in the previous experiments were far from those obtained here, where it was possible to clearly visualize the labeling of the meristem area, distinct from the uptake at the level of root hairs above the area of differentiation.

Figure 4.50 compares the root site where the phosphate transporter was expressed and the position where ^{32}P-phosphate was actually taken up. As clearly shown in the figure, phosphate was actually taken up from the site where the transporter gene was expressed. This imaging study was further developed to illustrate the importance of the role of the PHT1 family of phosphate transporters [20].

Finally, the resolution of the micro-RRIS is described. To investigate the resolution of the image, for example, a standard grid sample prepared by metals is commercially available for electron microscope imaging. However, standard grid samples prepared with radioisotopes (RIs) are not available. Therefore, the RI grid had to be prepared in our laboratory to estimate the resolution of the micro-images. The RI grid was prepared by printing RI lines with a printer where ^{137}Cs was mixed

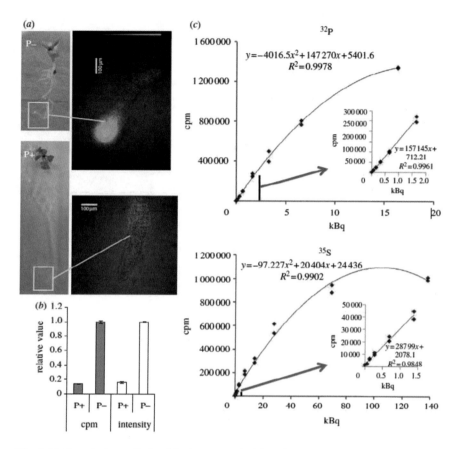

Fig. 4.48 Quantitative analysis of the image acquired by micro-RRIS [15]. (**a**) Arabidopsis was grown in sufficient or deficient conditions of phosphate for 10 days, ^{33}P-phosphate solution (6 kBq/ μL) was supplied for 5 min, and the root was imaged for 5 min. The radioisotope image of the root superposed on the light image showed a higher ^{33}P signal at the root tip when grown under phosphate-deficient conditions. (**b**) Measurements of plant radioactivity in the root tip (2 mm). The *gray* and *white* columns show the radioactivity counts of the root tip measured by a scintillation counter and image analysis, respectively. (**c**) Calibration curves of the counts of the image versus the radioactivity of the standard solution

with ink. The width of the line was 50 μm, and the distance between the lines was 450 μm (Fig. 4.51). From the image, the resolution of the micro-imaging system was estimated to be approximately 50 μm.

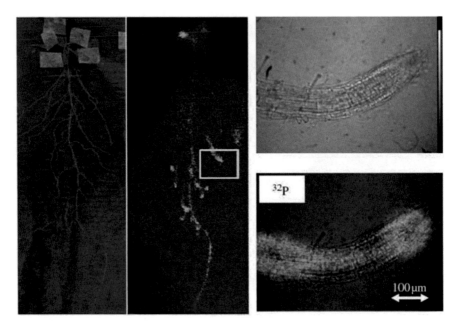

Fig. 4.49 ^{32}P uptake image of Arabidopsis root with higher magnitude [15] modified. ^{32}P-phosphate (100 Bq/µL) was applied to 10-day-old seedlings for 5 min, and the roots were imaged for 3 h. The isotope image was superposed on the corresponding light image, and pseudocolor was added according to the intensity of radioactivity. Using a modified micro-RRIS, a ^{32}P image was acquired at higher magnification with an optical lens (20 times)

4.5 Summary and Further Discussion

Real-time radioisotope (RI) imaging systems were developed step by step to image nutrient uptake in plants using not only conventional β-ray, γ-ray, or X-ray emitters but also additional produced radioisotopes, such as ^{28}Mg or ^{42}K. These methods provide specific and direct imaging possibilities for many ions where no alternative solution with fluorescent probes exists. The image allowed quantitative analysis to calculate the amount of the element and offered a wide dynamic range of detection. Therefore, it was possible to study the ion influx from culture medium over time periods as short as 1 or 2 min or over a period of a few days. This method also enabled us to conduct several types of pulse chase experiments, which cannot be performed using other tools, such as fluorescence techniques, which are mostly restricted to distribution analyses. The RIs we could test included over 10 nuclides in total. When the radiation energy is lower, the specific activity of the RIs should be higher; that is, a higher dose is necessary to acquire the image.

Through real-time imaging of the radioisotope movement, ion-specific speeds or accumulation patterns throughout the whole plant can be acquired. Among the elements investigated, there were three types of movement patterns in Arabidopsis.

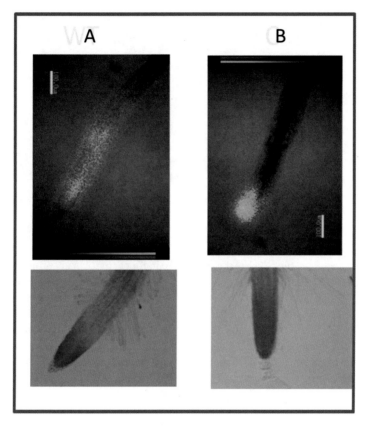

Fig. 4.50 ^{32}P image of Arabidopsis root, control, and mutant. (**a**) control sample; (**b**) mutant in which the phosphate transporter gene is expressed only at the root tip. *Upper*: ^{32}P images; *Lower*: gene expression images. There was a correlation between the gene expression area and the ^{32}P uptake area. *Bar*: 100μm

These ions are dissolved in water and therefore seem to move in the water but not with water. When the movement of water was measured, as described in Chap. 2, it showed constant speed and a constant route and expansion of the movement. On the other hand, the ion movement here showed specific speeds unrelated to the water movement itself.

This means that each element produces a specific concentration gradient from tissue to tissue. These profiles or movement might be different with time and change with the developmental stage. Thus, plants show very sophisticated and complicated element movement and profiles, and we do not have any means to understand the balance among all these movements.

Another interesting issue is how plants are absorbing elements. As was written regarding neutron imaging, many questions of root activity were raised when the movement of water or ions toward roots or within the plant was visualized. In this

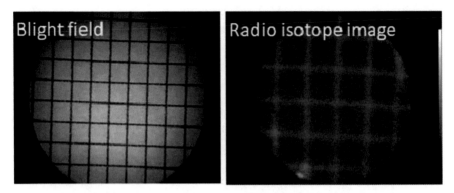

Fig. 4.51 Grid image of ^{32}P under micro-RRIS. To estimate the resolution of the image, ^{32}P was mixed with ink of the printer, and the grid pattern was drawn. The printed ^{32}P line width was 50μm at an interval of 500μm. Both bright field image (*left*) and ^{32}P image clearly reveal the printed lines

section, it was possible to show the difference in root activity in absorbing phosphate between water culture and soil culture.

Two types of real-time RI imaging systems (RRIS) have been presented, one for macroscopic imaging and the other for microscopic imaging of the plants treated with RIs. One of the purposes of macroscopic imaging is to pursue the possibility of tracing RIs without any influence on growth conditions, including illumination, and to image a wide range of the elements that contribute to plant growth. In the case of microscopic imaging systems, further development of the devices is needed. For example, developing a small growth chamber for plants with nutrient circulation and an irradiation system is now another goal. The resolution of microscopic imaging was approximately 50–100 μm.

It should be noted that additional equipment mounted on the microscope is needed to proceed with the acquisition of light signals, fluorescence or luminescence, at the same time, which could be combined or superposed with radioactive imaging. This system offers a broader range of applications, such as measuring the effects of the genetic manipulation of transporters labeled with green fluorescent protein or luciferase on ion transport in specific plant tissues.

The development of these macroscopic and microscopic imaging systems will facilitate the systematic analysis of the real-time uptake of various macro- and micronutrients, from the macroscopic level to the microscopic level, i.e., from the whole plant to the cellular level. Because these visualizations allow numerical analysis of the image, we expect that isotope images will open a new avenue in plant physiology.

Bibliography

1. Masumori M, Nogawa N, Sugiura S, Tange T (2016) Radiocesium in timber of Japanese cedar and Japanese red pione in the forests of MInamisoma, Fukushima. In: Nakanishi TM, Tanoi K (eds) Agricultural implications of Fukushima nuclear accident, pp 161–174
2. Sugita R, Hirose A, Kobayashi NI, Tanoi K, Nakanishi TM (2016) Imaging techniques for radiocesium in soil and plants. In: Nakanishi TM, Tanoi K (eds) Agricultural implications of Fukushima nuclear accident, pp 247–263
3. Sugita R (2014) Ph.D. thesis. The University of Tokyo
4. Sugita R, Kobayashi NI, Hirose A, Tanoi K, Nakanishi TM (2014) Evaluation of in vivo detection properties of ^{22}Na, ^{65}Zn, ^{86}Rb, ^{109}Cd and ^{137}Cs in plant tissues using Ral-time radioisotope imaging system. Phys Med Biol 59:837–851
5. Rai H, Kanno S, Hayashi Y, Ohya T, Nihei N, Nakanishi TM (2008) Development of a real-time autoradiography system to analyze the movement of the compounds labeled with beta-ray emitting nuclide in a living plant. Radioisotopes 57:287–294
6. Kanno S (2010) Ph.D. thesis. The University of Tokyo
7. Nakanishi TM (2017) Research with radiation and radioisotopes to better understand plant physiology and agricultural consequences of radioactive contamination from the Fukushima Daiichi nuclear accident. J Radioanal Nucl Chem 311:947–971
8. Kanno S, Rai H, Ohya T, Hayashi Y, Tanoi K, Nakanishi TM (2007) Real-time imaging of radioisotope labeled compounds in a living plant. J Radioanal Nucl Chem 272:565–570
9. Nihei N (2010) Ph.D. thesis. The University of Tokyo
10. Yamawaki M, Kanno S, Ishibashi H, Noda A, Hirose A, Tanoi K, Nakanishi TM (2009) The development of real-time RI imaging system for plant under light environment. J Radioanal Nucl Chem 282:275–279
11. Sugita R, Kobayashi NI, Hirose A, Iwata R, Suzuki H, Tanoi K, Nakanishi TM (2017) Visualization of how light changes affect ion movement in rice plants using a real-time radioisotope imaging system. J Radioanal Nucl Chem 312:717–723
12. Sugita, Kobayashi NI, Hirose A, Tanoi K, Nakanishi TM (2019) Visualization of ion transport in plants. In: Nakanishi TM, O'Brian M, Tanoi K (eds) Agricultural implications of Fukushima nuclear accident, pp 221–231
13. Hirose A, Yamawaki M, Kanno S, Igarashi S, Sugita R, Ohmae Y, Tanoi K, Nakanishi TM (2013) Development of a ^{14}C detectable real-time radioisotope imaging system for plants under intermittent light environment. J Radioanal Nucl Chem 296:417–422
14. Sugahara K, Sugita R, N Kobayashi NI, Hirose A, Nakanishi TM, Furuta E, Sensui M, Tanoi K (2019) Plastic scintillators enable life imaging of ^{32}P-labeled phosphorus movement in large plants. Radioisotopes 68:73–82
15. Kanno S, Yamawaki M, Ishibashi H, Kobayashi NI, Hirose A, Tanoi K, Nussaume L, Nakanishi TM (2012) Development of real-time radioisotope imaging systems for plant nutrient uptake studies. Philos Trans R Soc B 367:1501–1508
16. Kobayashi NI (2013) Time-course analysis of radiocesium uptake and translocation in rice by radioisotope imaging. In: Nakanishi TM, Tanoi K (eds) Agricultural implications of Fukushima nuclear accident, pp 37–48
17. Sugita R, Kobayashi NI, Hirose A, Saito T, Iwata R, Tanoi K, Nakanishi TM (2016) Visualization of uptake of mineral elements and the dynamics of photosynthates in Arabidopsis by newly developed real-time radioisotope imaging system (RRIS). Plant Cell and Physiology 57:743–753
18. Kobayashi NI, Ogura T, Takagi K, Sugita R, Suzuki H, Iwata R, Nakanishi TM, Tanoi K (2018) Magnesium deficiency damages the youngest mature leaf in rice through tissue-specific iron toxicity. Plant Soil 428:137–152

19. Rai H, Kanno S, Hayashi Y, Nihei N, Nakanishi TM (2008) Development of a fluorescent microscope combined with a real-time autoradiography system. Radioisotopes 57:355–360
20. Kanno S, Arrighi J-F, Chiarenza S, Bayle V, Berthome R, Peret B, Javot H, Delannoy E, Marin E, Nakanishi TM, Thibaud M-C, Nussaume L (2016) A novel role for the root cap in phosphate uptake and homeostasis. elife 5:e14577

Chapter 5
Visualization of ^{14}C-labeled Gas Fixation in a Plant

Keywords ^{14}C · ^{14}CO$_2$ · ^{14}CO$_2$ gas supply · Real-time CO$_2$ gas fixation image · Photosynthate movement image · Photosynthate transfer route · Phloem partition · Rice · Corn · Arabidopsis · Large-scale imaging

5.1 Performance of RRIS for ^{14}C imaging

To investigate the performance of RRIS for ^{14}C imaging, standard sources of ^{14}C were prepared by spotting ^{14}C-labeled sucrose on a polystyrene sheet from 1.8 Bq (0.5 Bq/mm^2) to 14,800 Bq (4500 Bq/mm^2). The spots on the sheet were measured at integration times of 3–60 min. Then, the minimum limit and upper limit of determination were evaluated in the RRIS (Real-time RI Imaging System) as well as the IP using ^{14}C standard spots, in the same way described for the development of RRIS (Chap. 4). The minimum limit of determination in the RRIS was lower than that in the IP when the accumulation time was relatively short. The linearity between the activity of ^{14}C and the signal intensity obtained by the RRIS was evaluated using the standards. There was good linearity between the activity count of ^{14}C and the signal intensity obtained by both the RRIS and an IP (data not shown). The minimum and upper limits as well as the dynamic range of determination measured for both the RRIS and the IP method are listed in Table 5.1.

The next step was to compare the image of a plant sample, supplied with ^{14}C-labeled CO$_2$ gas, taken by RRIS with that taken by an IP. The sample used for imaging was a seedling of *Arabidopsis thaliana* (Col-0) grown in culture solution for 38 days with flowers and developed pods. ^{14}C-labeled CO$_2$ was produced by mixing ^{14}C-labeled sodium hydrogen carbonate (2–5 MBq) with lactic acid in a 1.5 mL vial with a septum cap equipped with a syringe needle. The plant was placed in a polypropylene bag with the mouth sealed with clay in which a tube was connected. Then, 5 MBq of ^{14}CO$_2$ gas, generated in the vial, was introduced to the bag through a syringe for 24 h (Fig. 5.1). To fix the sample to an FOS where Cs (Tl)I scintillator was deposited, a silicone gum sheet was used, and the FOS was covered with polyphenylene sulfide film to prevent contamination with ^{14}C.

T. M. Nakanishi, *Novel Plant Imaging and Analysis*,
https://doi.org/10.1007/978-981-33-4992-6_5

Table 5.1 Properties of the RRIS and IP for ^{14}C imaging

	3 min	5 min	10 min	15 min	30 min	60 min
RRIS						
The minimum limit (Bq/mm^2)	4	2	2	1	1	1
The upper limit (Bq/mm^2)	1×10^3	1×10^3	1×10^3	1×10^3	1×10^3	1×10^3
Determined for the range	2×10^2	5×10^2	5×10^2	1×10^3	1×10^3	1×10^3
IP						
The minimum limit (Bq/mm^2)	8	4	2	2	1	1
The upper limit (Bq/mm^2)	4×10^3	4×10^3	2×10^3	1×10^3	2×10^2	2×10^2
Determined for the range	5×10^2	1×10^3	1×10^3	5×10^2	2×10^2	5×10^2

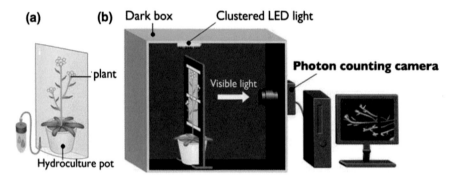

Fig. 5.1 Imaging of the ^{14}CO$_2$ gas fixation by the RRIS [1]. (**a**) ^{14}CO$_2$ gas was produced by mixing ^{14}C-labeled sodium hydrogen carbonate (2–5 MBq) with lactic acid and supplied to a polytropylene bag covering the plant sample. (**b**) Outline of the RRIS. A plant was prepared on the FOS in a dark box, and the image was monitored by a computer

First, images of the plant supplied with ^{14}CO$_2$ gas were taken by both the RRIS and an IP, and the images were compared. The exposure times were equal for the RRIS and IP. The images differed in quality, such as resolution, data depth, and contrast, particularly at sites with complex morphology such as a flower (Fig. 5.2). The main reason for the differences in image quality between the RRIS and IP was derived from the resolution. The resolution of the images captured by the RRIS was approximately 70 dpi, while that of the images captured by the IP was 500 dpi, which was more than seven times higher than that of the RRIS. In addition, the fixation method of the plant also caused differences in the quality of the images. The plant was fixed closely on the IP in a cassette for exposure, whereas in the case of RRIS, the plant was fixed relatively loosely to the FOSs using tape to allow growth during imaging. In total, the quality of the IP image was higher than that of the RRIS image, particularly at sites of complex morphology such as a flower.

The β-ray energy emitted from ^{14}C is low; hence, the effect of self-absorption cannot be ignored. Since the main factor in self-absorption is the thickness of the

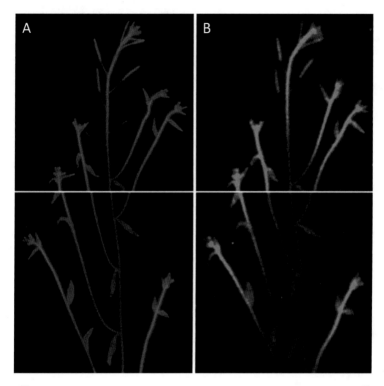

Fig. 5.2 ^{14}C images of an Arabidopsis acquired by an IP and RRIS [1]. Five MBq of $^{14}CO_2$ gas was supplied to the plant for 24 h after 38 days of germination. A and B are the images by an IP and RRIS with resolution of approximately 500 and 70 dpi, respectively. One of the reasons for the high resolution of the IP image was that the plant could be tightly fixed on the IP in a cassette for exposure

sample, the effect of self-absorption on the efficiency of signal detection by the RRIS was investigated before the analysis of the image. The radioactivity of the different tissues was measured, and the β-ray counts were compared with the intensity of the images. After 24 h of $^{14}CO_2$ supply to the plant, plant tissues such as the leaves, siliques, stems, and flowers were separated, and the radioactivity of ^{14}C in each tissue was counted for 2 min by a liquid scintillation counter (ALOKA LSC-6100). For the quenching effect in the counting, a known activity of ^{14}C was added to each sample as an internal standard, and the counts were corrected.

   Figure 5.3 shows the relationship between the signal intensity acquired by the RRIS and the actual radioactivity measured. The result shown in the figure indicates that the ^{14}C signal intensity found in some tissues cannot be linear. Thick tissues, such as mature leaves, flowers, siliques, and stems, showed no correlation between ^{14}C activity and the intensity of the image. These results indicated that it was difficult to quantify the ^{14}C activity in a thick sample using the autoradiographic technique in both imaging methods, IP and RRIS, because of self-absorption. On the other hand, there was potential for quantifying the ^{14}C activity in young and mature leaves using

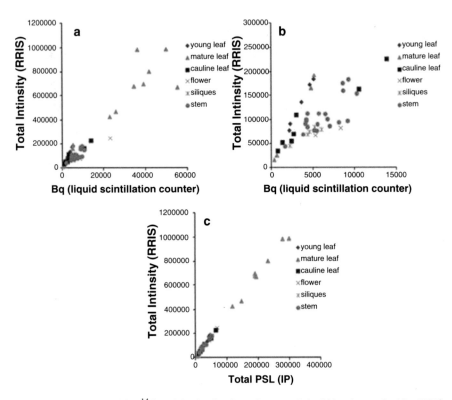

Fig. 5.3 Calibration of the ^{14}C activity in the tissue images of Arabidopsis acquired by RRIS [1]. (**a**) ^{14}C intensity versus radioactivity counts; (**b**) Magnification of the lower counts in a; c: relation between PSL value obtained by an IP and the intensity of the image obtained by RRIS

the RRIS. When we compared the intensity of the RRIS with the PSL value of the IP, linearity was maintained in most of the tissues: mature leaves, flowers, siliques, and stems. The results showed that although self-absorption was confirmed, the images taken by RRIS and IP seemed to show the same images of the ^{14}C profile, indicating that the IP image could compensate for the RRIS image.

As an example of ^{14}C imaging other than ^{14}CO$_2$ gas fixation, ^{14}C images of Arabidopsis are briefly introduced. In this case, ^{14}C-labeled sucrose, an initial metabolic product of photosynthesis, was supplied from the leaves as a kind of foliar fertilization. Figure 5.4 shows successive RRIS images of ^{14}C in a 30-day-old Arabidopsis plant supplied with ^{14}C-labeled sucrose from the rosette of the plant (1.85 MBq/500 μL) during the measurement. Although self-absorption had to be taken into account, the tendency of ^{14}C accumulation over time could be observed, since when the imaging area in the tissue was fixed, the change in the ^{14}C signal in the area indicated the relative change in the sucrose amount. Then, IP images of the whole plant at this stage and further aged samples were taken. Figure 5.5 is an IP image of the Arabidopsis plant at different stages of growth after 30 and 60 days of

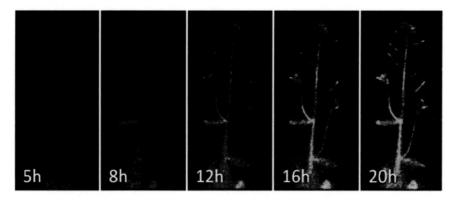

Fig. 5.4 An example of ^{14}C-sucrose images of Arabidopsis (1) [2]. 1.85 MBq/500 μL of ^{14}C-sucrose solution was applied to a rosette leaf of Arabidopsis after 30 days of germination. Successive ^{14}C images of the plant were acquired by RRIS. The integration time was 15 min every hour. The light was turned off during the image acquisition

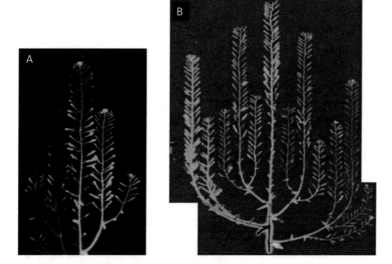

Fig. 5.5 Example of ^{14}C-sucrose images of Arabidopsis (2) [2]. 1.85 MBq/500 μL of ^{14}C-sucrose solution was supplied to rosette leaves of Arabidopsis after 30 (A) and 60 days (B) of germination

culture. Both figures showed the tendency of ^{14}C accumulation, such as that the sucrose assimilation products accumulated in maturing tissue and in joint parts. Since it was difficult to discern the detailed distribution of the silique, the pods were harvested, and the distribution of ^{14}C was investigated. The young pods, after 2–3 days of flowering, showed uniform accumulation of ^{14}C in the longitudinal direction by IP. However, a higher accumulation of ^{14}C was observed by IP at a higher position in the pod after 7 days of flowering (data not shown).

5.2   Imaging the $^{14}CO_2$ Gas fixation

After $^{14}CO_2$ gas is fixed by photosynthesis, the metabolites move to other tissues by phloem flow. This means that when the tissue to apply $^{14}CO_2$ gas for fixation is selected and the movement of photosynthate is traced, the sink-source relationship between the tissues can be analyzed. To study this relationship, $^{14}CO_2$ gas fixation imaging was performed.

5.2.1   Imaging of 43-Day-Old Plant

To visualize the flow of photosynthate, $^{14}CO_2$ gas was supplied to the rosette leaves of Arabidopsis after 43 days of growth. The rosette leaves were covered with polyethylene bags 1.2 μm in thickness. $^{14}CO_2$ gas was introduced to the bag for 24 h under light irradiation in a phytotron. Then, the upward movement of ^{14}C-labeled photosynthate was visualized by RRIS.

Figure 5.6 shows the change in the signal intensity during 24 h of $^{14}CO_2$ gas supply. The amount of ^{14}C-labeled metabolite in the main stem was very low, and hardly any signal appeared in the upper part of the main stem. In contrast, at the tip of the lateral stem, the amount of ^{14}C-labeled metabolite continued to accumulate, suggesting that the rosette leaves were the source organs supplying photosynthates primarily to the lateral stems. This result was surprising: despite the sink tissue present in the main stem, such as developed flowers and siliques, the main part of the sink was in the secondary stem. However, in these images, the phloem flow along the sieve tube connecting the basal shoot and the tip region around the main stem was not observed.

The plant sample tested was well developed, with mature siliques in the main stem and the lateral stem developing younger tissue than the main stem. Since the

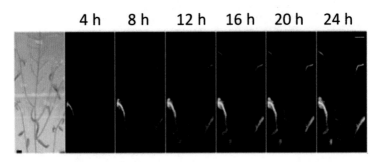

Fig. 5.6 Successive images of ^{14}C photosynthates after fixation of $^{14}CO_2$ gas in Arabidopsis [3]. (Photograph of the plant and successive images of the ^{14}C-labeled metabolite movement. $^{14}CO_2$ was supplied to rosette leaves of 43-day-old plants, and the upward movement of ^{14}C-phtosynthates was visualized by RRIS. The accumulation time of the imaging was set as 15 min. Scale bar: 2 cm)

primary photosynthates preferentially accumulated at the lateral stem, it was hypothesized that the rosette leaves are the source organs when the stem is young, but after flowering, the necessary carbon source in the stem is supplied by photosynthates produced in the siliques, stems, and cauline leaves. To test this hypothesis, the same experiment was performed using a younger plant.

5.2.2   Younger Sample Imaging

To test the hypothesis mentioned above, the same experiment was performed using a younger plant, 30 days after germination. Accordingly, the movement of the photosynthates was different from that in the 43-day-old plants. When $^{14}CO_2$ gas was introduced to the rosette leaves, the photosynthate produced in the rosette leaves was preferentially transferred to the main stem tip (Fig. 5.7). The direction of phloem flow from the rosette leaves towards each stem was changed in the basal shoot region and was influenced by the age of the stem.

In contrast, there was no difference in the amounts of ^{14}C detected at the tips of the main and lateral stems when $^{14}CO_2$ gas was supplied to the whole shoots. However, when $^{14}CO_2$ gas was supplied to the aboveground part of the plant, except for the rosette leaves, the ^{14}C intensity was higher at the lateral stem than at the main stem, an opposite movement from that supplied from the rosette (Fig. 5.7). This observation suggested the existence of a source organ other than the rosette leaves that supplied photosynthates to the lateral stems.

To determine whether the potential source organ to the lateral stem was the inflorescence, $^{14}CO_2$ gas was supplied in a pulse of 1 h to the inflorescence only. The result was that the ^{14}C signal intensity in the lateral stem tip was much higher than that in the main stem tip, and this high intensity of the ^{14}C signal in the lateral

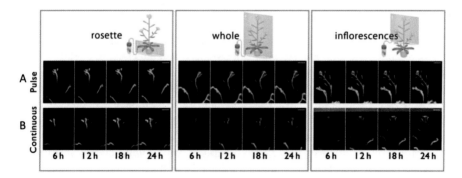

Fig. 5.7 ^{14}C-photosynthate movement in Arabidopsis when $^{14}CO_2$ gas was supplied to different tissues [3]. Successive images of ^{14}C photosynthate are shown after/while supplying $^{14}CO_2$ to rosette leaves, whole plant or inflorescence. Serial images were acquired by RRIS. $^{14}CO_2$ gas was supplied as pulses (A) or continuously (B). The imaging time for each frame was 15 min. Scale bar: 20 mm

tips continued up to 24 h, suggesting that the ^{14}C metabolites produced in the main stem are continuously transported towards the lateral stems. In the case of the photosynthates produced in the rosette leaves, preferential transfer to the main stem tip was observed after a pulse supply of ^{14}CO$_2$ gas, similar to that resulting from the continuous supply.

5.2.3 *Photosynthate Transfer Route by Image Analysis*

To analyze the photosynthate transfer route, the changes in the ^{14}C signal in the main stem tip and lateral stem tip were plotted. During the continuous ^{14}CO$_2$ gas supply, the orientation of the photosynthate movement showed that there was a difference in the route as well as in the photosynthate accumulating tissue according to the gas fixation site, the rosette or the aboveground part without the rosette.

As shown in Fig. 5.8, the preferential transfer of photosynthate to the main stem tip was observed when ^{14}CO$_2$ gas was supplied from the rosette for 24 h. However, the ^{14}C accumulation at the lateral stem tip linearly increased during 24 h of continuous ^{14}CO$_2$ supply from the inflorescences, whereas the rate of increase at the main stem tip was lower than that of the lateral stem after approximately 12 h. In the case of ^{14}CO$_2$ gas supplied from the whole plant, the increase in the ^{14}C signal was similar between the main stem tip and lateral stem tip.

When ^{14}CO$_2$ gas was supplied for 1 h as a pulse, the ^{14}C signal increase was similar to that when ^{14}CO$_2$ gas was supplied continuously, except for the supply from the inflorescences. The photosynthate supply clearly moved towards the lateral stem tip, where the accumulated amount of photosynthates was more than two times higher than that in the main stem tip. The high ^{14}C signal intensity kept in the lateral stems by the pulse supply and the steady increase during the continuous supply from the inflorescences suggested that ^{14}C-labeled photosynthates generated in the main stem, other than the rosette leaves, are continuously transported to the lateral stems. Although the route of photosynthate movement was different depending on the production site of ^{14}CO$_2$ gas fixation, in total, there was no significant difference in the accumulation amount among plant tissues in a younger plant, as shown when the gas was supplied to the whole plant.

To analyze the route of photosynthates after production in the aboveground part of the plant except for the rosette leaves in more detail, cauline leaves and tips in the lateral stem were plotted. The ^{14}C signal in the cauline leaves of the lateral stems was decreased, although the total signal intensity of ^{14}C in the cauline leaves and lateral stem tips was maintained (Fig. 5.8c). This observation might indicate that cauline leaves also act as a carbon source for lateral stem tips, although the amount of ^{14}C-photosynthate in the lateral stem tip area, including the tips and cauline leaves, was kept constant.

The next analysis was to determine which tissue producing photosynthate provided a carbon source to the silique. The region of interest (ROI) was set as shown in Fig. 5.9, and siliques were numbered Si1 to Si4, from the lower to the upper part of

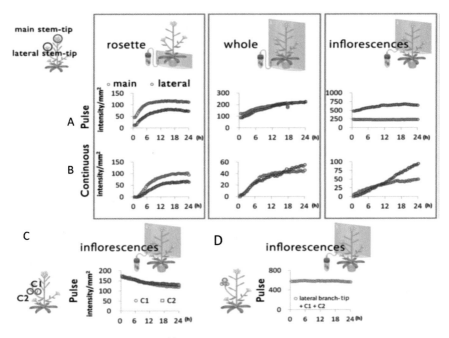

Fig. 5.8 Preferential route of the ^{14}C-photosynthate movement in the main or lateral stem in Arabidopsis [3]. $^{14}CO_2$ gas was supplied under both pulsed (A) and continuous (B) conditions to rosette leaves, whole plants or inflorescences. ROIs were set as the main (blue) and lateral (red) stem tips for A and B. Two ROIs were set on the cauline leaves (C1 and C2) for C, and three ROIs were set for the lateral stem. D: Two cauline leaves and the lateral stem tip. The preferential transfer of ^{14}C-photosynthate to the main stem tip was shown when $^{14}CO_2$ gas was supplied to the rosette leaves, whereas the preferential move to branch tip was shown when the gas was supplied inflorescence

the plant. When $^{14}CO_2$ gas was supplied to the rosette leaves as a pulse for 1 h, the ^{14}C amounts in all the siliques increased. The ^{14}C signal also increased according to the position of the silique from low to high, i.e. the younger siliques accumulated higher amounts. This result indicated that the C source in the siliques was derived from rosette leaves. On the other hand, when $^{14}CO_2$ gas was supplied to the inflorescence as a pulse, the ^{14}C amount decreased with time in all the siliques, and there was hardly any difference in amount among the siliques. This result, that the ^{14}C amount in the siliques was slightly decreased when the photosynthate from rosette leaves contained ^{14}C, suggested that there was hardly any movement of the photosynthate from the siliques to the other tissues and that the loss of ^{14}C at the silique might be due to respiration. A pulse supply of $^{14}CO_2$ gas to the whole plant showed that even when photosynthate was supplied from the rosette leaves, the amount of ^{14}C decreased in older siliques. With a continuous supply of $^{14}CO_2$ gas to the inflorescences, the ^{14}C amount increased in all the siliques measured. This result indicated that photosynthate from the main stem, including the cauline leaves or siliques, could be supplied to younger siliques.

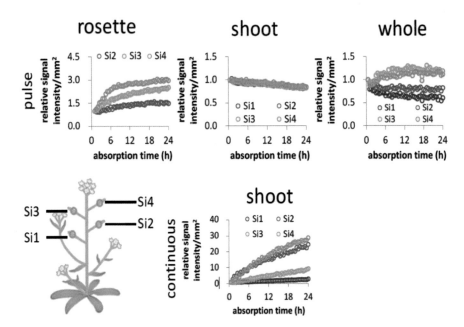

Fig. 5.9 Preferential route of the ^{14}C-photosynthate movement to silique in Arabidopsis [4]. [14]CO$_2$ gas was supplied as a pulse (1 h) to rosette leaves, whole plants or shoots (inflorescences). In the case of shoots, [14]CO$_2$ gas was continuously supplied. ROIs were chronologically set at siliquis in the main stem from Si1 to Si4. There was a preference for the ^{14}C-photosynthate transfer route when supplied from rosettes

The method of photosynthate transfer could be discussed in comparison with rape seed plants, whose rosette leaves fall down after flowering with increasing photosynthesis activities of cauline leaves. Although the rosette leaves of Arabidopsis do not fall down after flowering, the photosynthesis activity gradually decreases after maximum expansion of rosette leaf development, and even after senescence of the rosette leaves, the emergence of siliques continues for a while suggesting the important role of photosynthate produced in the cauline leaves or siliques in Arabidopsis, similar to that in rape seed plants.

5.2.4   Whole Plant Image of Photosynthate by an IP

To obtain ^{14}C images of the whole plant, including roots, the plants were harvested and placed on an IP after imaging by RRIS. Figure 5.10 shows the autoradiograph of the plant after [14]CO$_2$ gas was supplied for 1 h. As shown in the figure, the profile of the ^{14}C signal in the aboveground part of the plant showed the same profile as that obtained by RRIS. The root image acquired by the IP, which was not obtained in the RRIS, also indicated the difference in the ^{14}C signal with respect to the difference in

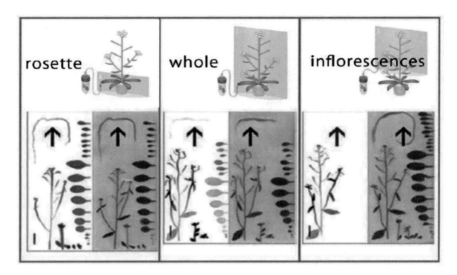

Fig. 5.10 Preferential route of the ^{14}C-labeled metabolite transfer to the root in Arabidopsis [3]. The ^{14}C-labeled metabolite distribution images by IP were acquired after imaging by RRIS, where $^{14}CO_2$ gas was supplied as pulses (1 h) to rosette, whole plant or inflorescences. In each $^{14}CO_2$ gas supply method, the grayscale image by an IP (*left*) and photograph (*right*) are shown. Scale bar: 20 mm. The arrows indicate the roots. When $^{14}CO_2$ gas was supplied to the rosette, ^{14}C-labeled metabolites were preferentially transferred to the root, whereas ^{14}C-labeled metabolites were not observed in the root (*blue arrow*) when the gas was supplied from inflorescences

the $^{14}CO_2$ gas supply site, from rosette leaves or inflorescences. The red and blue arrows in Fig. 5.10 indicate the root part in the picture and in the radiograph, respectively. It was noted that no ^{14}C image of the roots was observed when the $^{14}CO_2$ gas was supplied only to the aboveground part of the plant other than the rosette leaves, as shown by the blue arrow in the root autograph. However, when $^{14}CO_2$ gas was supplied to the rosette leaves, a higher amount of ^{14}C was shown in the root than when the gas was supplied to the whole plant. The results indicated that rosettes are the primary carbon source for roots. Thus, the photosynthate fixed in the inflorescence seemed to be transported, metabolized, and accumulated only within the inflorescence itself.

5.3 Photosynthate Movement in Soybean Plants When $^{14}CO_2$ Was Supplied

Using the RRIS, the carbon dioxide gas fixation process, as well as the orientation of the photosynthate in Arabidopsis, was visualized by applying the developed method of ^{14}C-labeled gas supply. The next experiment was to visualize photosynthate movement by employing soybean plants. Since soybean is much larger than Arabidopsis, it was easier to supply $^{14}CO_2$ gas to specific tissues and to study how

(a)   **(b)**

Fig. 5.11 ^{14}C-photosynthate distribution in a soybean plant [5]. [14]CO$_2$ gas was prepared and supplied to a whole soybean plant for 30 min after 40 days of germination. Then, the plant was exposed to an IP (imaging plate) for 30 min. a: Picture of the plant; b: Radiograph of the image in the IP. The ^{14}C photosynthate was uniformly distributed throughout the expanded leaves; however, the ^{14}C signal was not observed in the roots

the photosynthate moved from one tissue to other tissues. For this purpose, soybean plants (*Glycine max.* cv, Enrei) were grown in culture solution under 16 h L/8 h D, light/dark conditions in a phytotron. First, [14]CO$_2$ gas was supplied as a pulse for 30 min to the whole plant 40 days after germination. The ^{14}C signal was distributed uniformly among the developed leaves, suggesting that a similar amount of photosynthate was produced by 30 min of photosynthesis in each leaf. Figure 5.11 is the image of the plant acquired by an IP. There was no ^{14}C signal observed in the roots, suggesting that most of the photosynthate produced in the leaves remained at the site where photosynthesis occurred; therefore, it was not transferred to the roots.

Since the amount of photosynthate is similar among the expanded trifoliate leaves, the first trifoliate leaves of 20-day-old plants were chosen, and [14]CO$_2$ gas was supplied only to these leaves for 30 min. Then, the photosynthate movement to other tissues was analyzed from the decrease in the ^{14}C signal from these leaves. Figure 5.12 shows the change in the ^{14}C signal during 8 h after the gas was supplied. It was shown that the photosynthate produced in the original trifoliate leaves gradually decreased with time and plateaued after approximately 4 h. The decreasing

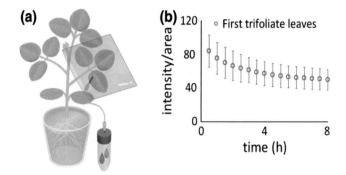

Fig. 5.12 ^{14}C-photosynthate transfer manner in the leaves of a soybean plant [5]. $^{14}CO_2$ gas was prepared and supplied to the trifoliate leaves for 30 min; then, real-time imaging (RRIS) was performed. The radioactivity intensity of the trifoliate leaves was monitored from the intensity of the sequential images taken by RRIS. a: Schematic illustration of the $^{14}CO_2$ gas supply to the trifoliate leaves; b Relative radioactivity intensity curve of the trifoliate leaves with time

curve of the relative intensity in the treated leaves showed that within 8 h, most of the transfer movement of the photosynthate in the phloem seemed to cease.

Using plants at the same growth stage, $^{14}CO_2$ gas was supplied to the selected tissue for 30 min, and the movement of the ^{14}C image was observed by the RRIS until 8 h after the treatment. From the trifoliate leaves, the metabolites were transferred preferentially to the youngest tissues and accumulated there. However, when $^{14}CO_2$ gas was supplied only to the youngest leaves, including the meristem, the photosynthate produced in this tissue remained at this site and hardly moved to the other tissues (Fig. 5.13). This preferential movement of photosynthate from trifoliate leaves to the youngest tissue was shown, in different to the position where the trifoliate leaves developed (data not shown). There are many kinds of tissue at different developmental stages in one plant; therefore, it was suggested that premier importance was placed on promoting the growth of the youngest tissue by transferring photosynthate, which was the source to produce the structure of the plant.

The movement of photosynthate from trifoliate leaves to the younger tissue could be observed at an earlier time when $^{14}CO_2$ gas was supplied for 30 min. However, the movement of photosynthate from the aboveground part to the root was much slower than that to the younger tissue, as shown above. The ^{14}C signal was not shown in roots by a 30-minute supply of $^{14}CO_2$ gas (Fig. 5.11). Since it took a longer time to move and accumulate the photosynthate in roots, Fig. 5.14 shows an image of the whole plant taken by an IP after 24 h of continuous supply of $^{14}CO_2$ gas. Since the leaves supplied with $^{14}CO_2$ gas emit higher radiation than the other tissues due to the remaining ^{14}C-labeled compounds, the background level of the image was increased when the whole plant was exposed to the same IP. Therefore, the treated leaves were disconnected from the plant, and the images of the cut off leaves and the rest of the plant were taken by different IPs. The original growing site of these trifoliate leaves is indicated by an arrow in the picture. The IP images showed that

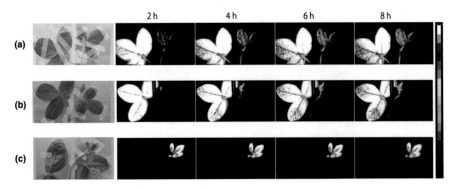

Fig. 5.13 ^{14}C-photosynthate movement in a soybean plant [5]. The [14]CO$_2$ gas was supplied to the trifoliate leaves of the plant after 20 days of germination for 30 min, and the movement of the photosynthate was monitored by RRIS. Sequential images are shown after [14]CO$_2$ gas was supplied. [14]CO$_2$ gas was supplied to the first (a, b) and third (c) trifoliate leaves. The accumulation time for each imaging frame was set to 15 min. The signal intensity is assigned using pseudocolor (white represents the highest intensity)

when [14]CO$_2$ gas was supplied to the expanded trifoliate leaves, most of the photosynthate was preferentially moved to the youngest tissue, and only a small amount of the photosynthate was moved to the roots or to other tissues, regardless of the position of the trifoliate leaves. The route preference was the same as that shown by RRIS. However, when [14]CO$_2$ gas was supplied to the youngest tissue, all of the photosynthate produced at this site remained at the site, and movement to other tissues, including roots, was not observed. Photosynthate transfer from the youngest tissue was hardly detected.

Another question was from what source photosynthate is supplied to the pods. To determine the orientation of the photosynthate movement from leaves to developing pods, plants after 55 days of germination were selected. Then, [14]CO$_2$ gas was supplied for 30 min to the trifoliate leaves grown at the site close to the pod. The accumulation images obtained from the RRIS and the IP are shown in Fig. 5.15. In the case of the pods, carbon fixed from the [14]CO$_2$ gas in the matured trifoliate leaves was preferentially transferred to the closest pod, and the accumulated amount was still increasing even after 8 h of [14]CO$_2$ gas supply. Although several pods developed on the plant, the accumulation of photosynthate was not observed in those developed at higher positions than the treated trifoliate leaves, although they were younger than the closest pod. However, a small amount of photosynthate accumulation was shown in the pods grown below the tissue where the [14]CO$_2$ gas was supplied, in leaves as well as in roots.

It was interesting to note that in the growth stage of pod development, the main carbon source, photosynthate, produced in expanded trifoliate leaves was not supplied primarily to the youngest tissues with meristems but to the closest pod. It was suggested that most photosynthate was not transferred for a long distance when pods were developing.

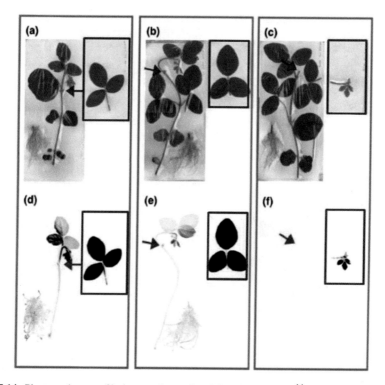

Fig. 5.14 Photosynthate profile in a soybean plant [5]. After 24 h of $^{14}CO_2$ gas supply to the trifoliate leaves and the youngest tissue, the plant was exposed to an IP. The upper figures in a, b, and c are pictures of the plant corresponding to each IP image below (d, e, and f). $^{14}CO_2$ gas was supplied to the expanded trifoliate leaves (a, b, d, and e) and youngest tissue (c and f). To reduce the background noise of the IP images, the treated leaves were disconnected from the plant and exposed to different IPs (d–f), and the positions where the leaves were connected are shown by arrows

5.4   Downward Movement of Photosynthate to Roots

When $^{14}CO_2$ gas was supplied to the rosette leaves, ^{14}C-labeled photosynthates were found to be transported to the root (Fig. 5.10). To visualize the sink tissues within the roots, imaging of the downward movement of ^{14}C-labeled photosynthates was performed after $^{14}CO_2$ gas was supplied only to the aboveground part of 14-day-old Arabidopsis seedlings, which are juvenile plants before flowering. Plants were grown in a 0.4% gellan gum and full-nutrient culture solution using a dish prepared with several vent holes. Plant roots were then placed on gellan gum on a polyethylene sheet (thickness: 10 μm) for imaging. The plants on the FOS were placed vertically, and images were acquired for 15 min at intervals of 1 h. The plants were irradiated by light-emitting diode light (100 μmol/m^2/s) for 45 min between the image acquisition periods.

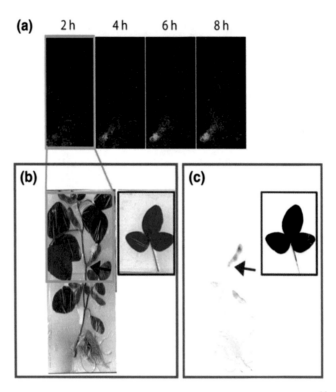

Fig. 5.15 Photosynthate movement from leaves to a pod of a soybean plant [5]. [14]CO_2 gas was supplied to the expanded trifoliate leaves of a soybean plant after 55 days of germination, and images of the ^{14}C-photosynthate movement with time were acquired by RRIS. After sequential images were taken, the plant was exposed to an IP. To decrease the background noise, the treated tissues were disconnected and exposed to different IPs. The disconnected site is shown by arrows (b and c)

The RRIS images of ^{14}C in roots visualized the arrival of ^{14}C-labeled photosynthates at the root tip areas, including developing lateral roots, as early as 3 h after [14]CO_2 was supplied. Thereafter, the accumulation of ^{14}C-labeled photosynthates in the root tips increased for 12 h. Then, the accumulation of ^{14}C-labeled photosynthates at the main root tip was observed under micro-RRIS. The root elongation rate in 2-week-old Arabidopsis plants was 5.1 ± 0.4 (SD) mm over 12 h. Therefore, the root tip segments shown in Fig. 5.16 under micro-RRIS were inferred to be newly developed tissues constructed with the ^{14}C-labeled photosynthate and thus were the sink of the photosynthates.

The result that [14] C-labeled photosynthates preferentially accumulated in the tip area is in agreement with the findings in *Brassica napus* seedlings, in which the photosynthate produced in leaves was translocated to the meristematic root regions (Dennis et al. 2010). The phloem unloading activity around the root tip of Arabidopsis has been previously visualized using carboxyfluorescein (CF) dye applied to a single cotyledon (Oparka et al. 1994). Based on sequential CF images

(a)   **(b)**   **(c)**

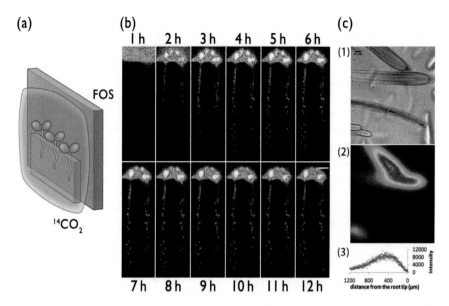

Fig. 5.16 Visualization of the downward movement of ^{14}C-labeled metabolites from leaves to roots [3]. (**a**) Schematic illustration of the sample covered with a polyethylene bag in preparation for the $^{14}CO_2$ gas supply. (**b**) Serial images of the ^{14}C-labeled metabolite movement taken by macro-RRIS. (**c**) After taking successive images by macro-RRIS, we obtained micro-RRIS images of ^{14}C-labeled metabolites in the root (root tip, maturation area and lateral root). (1): Grayscale image; (2): pseudocolor image; (3): distribution profile of ^{14}C-labeled metabolites in a root. Scale bar = 100 mm. The 14C signal was high at approximately 200 and 800 µm to the main root tip

taken by confocal laser scanning microscopy, the protophloem located 200–700 µm behind the root tip was suggested by the authors to function in phloem unloading and subsequent lateral transport. Consistent with this suggestion, a high ^{14}C signal intensity was detected approximately 200 and 800 µm distal to the main root tip using the micro-RRIS (Fig. 5.16). This region, now suggested to be the major sink tissue in roots, can be considered the part extending from the middle part of the apical meristem to the start of the elongation zone.

5.5 $^{14}CO_2$ Fixation in a Large-Scale Plant

The RRIS was developed using a fiber optic plate (FOS), on which a CsI (Tl) scintillator was deposited to convert radiation into light. However, the scintillator size was fixed at 10×10 cm, which was too small to observe the entire plant. Therefore, a plastic scintillator, Lumineard-C, was used to image the ^{14}C-labeled photosynthate movement in a plant (see Chap. 4, Sect. 4.2.6). The performance of the plastic scintillator was studied, and it was demonstrated that a plastic scintillator was applicable for ^{14}C imaging in a plant. To study the long-distance transportation of photosynthate, this system can visualize ^{14}C-labeled photosynthate movement between shoots and roots when the plant length is long.

A 40-day-old rice seedling (*Oryza sativa* L. cv. Nipponbare) and a 70-day-old maize seedling (*Zea mays* L.), a commercial hybrid sweet corn, were employed to analyze photosynthate movement in plants. They were grown in culture solution under 16 h L/8 h D, light/dark conditions. The aboveground heights of the rice and maize plants were approximately 550 mm and 400 mm, respectively. ^{14}CO$_2$ gas was produced by mixing ^{14}C-labeled sodium hydrogen carbonate and lactic acid in a 1.5 mL vial with a septum cap. Each plant was sealed with a polyethylene bag. Then, the vial and bag were connected using a tube to introduce ^{14}CO$_2$ gas into the bag. Doses of 4 MBq and 8 MBq were applied to rice and maize seedlings for 90 and 120 min, respectively. After the ^{14}CO$_2$ gas was supplied, the plant was removed from the bag, fixed to Lumineard-C (170 $\times$ 750 mm) covered with an Al sheet (2 μm in thickness), and placed in a large dark box. In the box, light was kept off for 15 min after a 15 min light period, and imaging was performed during the dark period. The imaging was continued until 24 h after the treatment.

Since ^{14}CO$_2$ gas was supplied for a limited time, the imaging showed how the ^{14}C metabolites moved after fixation. As expected, the amount of ^{14}C continuously decreased in both rice and maize leaves (Fig. 5.17). The pattern of decrease in the rice plant showed that the photosynthetic ability per unit area and the decreasing speed of photosynthate were approximately the same among the leaves. In the case of maize, older leaves showed a more rapid decrease in ^{14}C-labeled photosynthate, which seemed to be caused by the translocation of metabolites from leaves to other tissues, including roots, as well as loss from the tissue as ^{14}CO$_2$ gas by respiration. Since the photosynthate flow changes with the developmental stage, this result does not provide a reason to discuss the differences in the pattern of decrease in the photosynthate between the plants. However, the similar decrease in photosynthate among the leaves in the rice plant suggested that the rice plant itself was at a younger developmental stage than the maize plant.

5.6   Summary and Further Discussion

In the study on the performance of the RRIS imaging system, it was found to be possible to trace the movement of ^{14}C in plant tissue. This means that it was possible to trace the process of photosynthesis, carbon fixation, and photosynthate movement after ^{14}CO$_2$ gas was fixed in plant tissue. However, the imaging result had to be carefully analyzed because the effect of self-absorption of the β-rays emitted from ^{14}C could not be ignored. Considering all these conditions, the findings were as follows.

Visualizing the route of photosynthate movement in Arabidopsis after the assimilation of ^{14}CO$_2$ gas in plant tissue revealed a new finding about the flow. The images showed that the transfer route of the metabolites was dependent on the original tissue where the photosynthate was produced. Since the photosynthate was moving via the

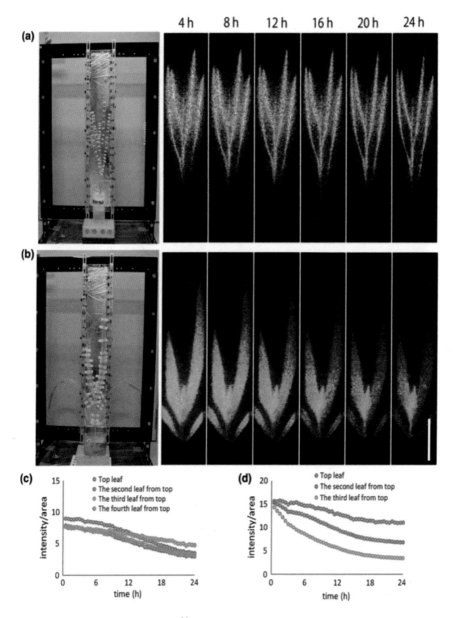

Fig. 5.17 Large-scale imaging of the ^{14}C-photosynthate movement in a plant [6]. (Photograph and sequential images of ^{14}C photosynthate in (a) 40-day-old rice and (b) 70-day-old maize using a Lumineard-C scintillator (170 × 750 mm). Scale bar: 100 mm. Time course of ^{14}C movement in (c) rice and (d) maize. The number of ROIs was set from new to old leaves)

phloem flow, it was also possible to trace the phloem flow by tracing the signal of ^{14}C. In addition to how the photosynthate was transferred, it was possible to analyze the phloem partition site and timing.

Notably, these source-sink relationships changed over time with the development of each tissue in the plant. In the case of Arabidopsis, it was shown, as we expected, that rosette leaves were the source organs when the stem was young, but after flowering, the necessary carbon source in the stem was supplied by photosynthates produced in the siliques, stems, and cauline leaves. This means that the role of the expanded leaves in supporting the other tissue changes with the development of the whole tissue of the plant.

The result presented above is one of the examples of analyzing photosynthate movement, but there is also much information in successive images; therefore, by setting a suitable ROI (region of interest) in the image, it was possible to analyze the movement of the photosynthate in more detail. It was amazing for us to be able to trace the photosynthate and define which tissue was created in the meristems from the fixed carbon visualized. However, to further analyze the newly created tissue and the dynamics of phloem unloading of photosynthates, micro-RRIS needs to be improved to supply the labeled gas under light conditions, especially for real-time imaging.

In the case of a soybean plant, high accumulation of photosynthate in the youngest tissue was visualized, similar to that of the other plants. When the photosynthate flow to the pod was visualized, although the amount of photosynthate production was at the same level among the leaves, the assimilated carbon was preferentially transferred from the trifoliate leaves to the closest pod. It is known that the pod is producing the optimal conditions for the growth of seeds, such as a high concentration of CO_2 gas and high photosynthesis activity within the pod. Though the self-absorption of the pod is high, with further development of the devices for the imaging system, the visualization of the CO_2 gas in a pod might be possible.

To visualize a larger sample, a plastic scintillator, Lumineard, was employed, and it was possible to visualize photosynthate movement in the leaves. As described earlier, although the self-absorption of ^{14}C, especially in the internode, is high, when the analyzed tissue was carefully chosen and the change in ^{14}C was carefully taken into account, it was possible to trace the photosynthate movement within the plant, especially plants growing in soil. In summary, it was shown that the detection of a gaseous radionuclide in macro- and micro-RRIS could drastically enhance the versatility of RRIS.

Bibliography

1. Sugita R, Kobayashi NI, Hirose A, Ohmae Y, Tanoi K, Nakanishi TM (2013) Nondestructive real-time radioisotope imaging system for visualizing ^{14}C-labeled chemicals supplied as CO_2 in plants using Arabidopsis thaliana. J Radioanal Nucl Chem 298(2):1411–1416

2. Ohmae Y, Hirose A, Sugita R, Tanoi K, Nakanishi TM (2013) Carbon-14 labelled sucrose transportation in an Arabidopsis thaliana using an imaging plate and real time imaging system. Jouranal of Radioanalytical and Nucl Chem 296:413–416
3. Sugita R, Kobayashi NI, Hirose A, Saito T, Iwata R, Tanoi K, Nakanishi TM (2016) Visualization of uptake of mineral elements and the dynamics of photosynthates in Arabidopsis by Newly developed real-time radioisotope imaging system (RRIS). Plant Cell and Physiology 57:743–753
4. Sugita R (2014) Ph.D. thesis. In: The University of Tokyo
5. Sugita R, Kobayashi NI, Tanoi K, Nakanishi TM (2018) Visualization of $^{14}CO_2$ gas fixation by plants. J Radioanal Nucl Chem 318:585–590
6. Sugita R, Sugahara K, Kobayashi NI, Hiroe A, Nakanishi TM, Furuta E, Sensui M, Tanoi K (2018) Evaluation of plastic scintillators for live imaging of ^{14}C-labeled photosynthate movement in plants. J Radioanal Nucl Chem 318:579–584

Chapter 6
3D Images

Keywords 3D image · ^{109}Cd · ^{137}Cs · Rice · Grain · Distribution · Accumulation · IP image

6.1 3D Image of ^{109}Cd in a Rice Grain

Seeds of rice plants (*Oryza sativa* L. var. Nipponbare) grown in water culture were employed for 3D image construction. During the maturing process of the grain, after flowering, ^{109}Cd (1 MBq) was supplied to the water culture. After 24 h of treatment, the brown rice was harvested and subsequently embedded in resin under freezing conditions. The rice grain was sliced into sequential sections 5 μm in thickness (Fig. 6.1), and the sliced sections were removed successively every 100 μm and stored at −20 °C. The height of the whole grain was approximately 6 mm; therefore, the number of slices was approximately 1200 sheets when sliced every 5 μm. Out of these 1200 sheets, successive sheets were removed every 100 μm, and these 60 sheets were used to construct the 3D image.

Since the ionic form of nuclides can move inside the plant during the sectioning process, such as during fixation and dehydration, sample preparation and radiography were performed under frozen conditions to diminish radionuclide movement. After the samples were placed on an IP, the image produced in the IP was read by a scanner to obtain the radiographs. Then, the successive ^{137}Cs distribution images acquired from the IP were used to construct the 3D image employing ImageJ software.

Figure 6.2a and c show the sliced grain picture and the radiation images of 2 sliced grains taken by an IP, respectively. In each unit (red square), two horizontal and two vertical images of the two grains are shown. In each unit, the two images on the left side and right side are the grains 20 and 15 days after flowering, respectively, showing a horizontal (small round image) and a vertical (longer image) pair of images for each grain. When the picture and the radiograph image are superposed (Fig. 6.2b), it was

Supplementary Information The online version of this chapter (https://doi.org/10.1007/978-981-33-4992-6_6) contains supplementary material, which is available to authorized users.

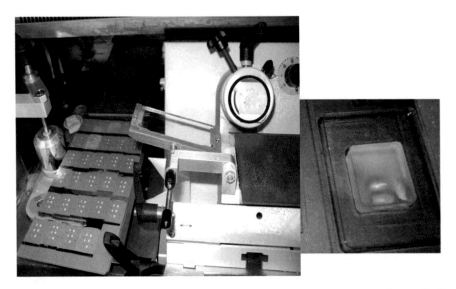

Fig. 6.1 Slicing the rice grain to construct 3D images. The grain was embedded in resin and sliced to 5 μm in thickness under freezing conditions

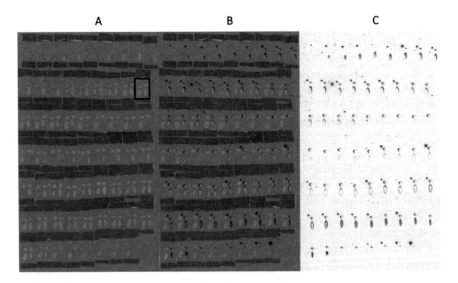

Fig. 6.2 Series of sliced grain images after 15 and 20 days of flowering [1] modified. For each small unit site, two pairs of slices were placed. The small and large slices in the unit are horizontally and vertically sliced grains. An example of the small site unit is shown in a *red square*. In each unit, two grains at each growth stage are shown: after 15 (right side) and 20 days (left side) of flowering. (**a**) Photograph of the sliced seeds; (**b**) radiograph image (**c**) superposed to a photograph of the sliced pictures (**a**); (**c**) ^{109}Cd radiograph image acquired by an IP

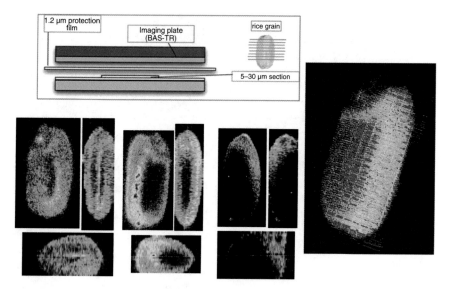

Fig. 6.3 3D image of [109]Cd in a rice grain constructed from the sliced image acquired by an IP [1] partially. In total, 62 images acquired by the IP were rearranged, and a 3D image was constructed. Upper: schematic illustration of the process to acquire the [109]Cd image in each slice; Lower right: 3D volume view; Lower left: orthogonal volume views. Pseudocolor was assigned to the radiation intensity

clear that there was hardly any [109]Cd observed in the left side grains in the unit, i.e., the grains 20 days after flowering.

The grains after 5, 10, 15, and 20 days were sliced in the same way, and from the series of approximately 60 images, 3D images were constructed. The volume of the endosperm increased from day 5 to 20 after flowering, along with the development of the grain. When 3D images were constructed, the decrease in [109]Cd accumulation at the endosperm was more clearly shown. However, the [109]Cd image of the grain 20 days after flowering was hardly observed. Figure 6.3 shows the 3D images of the grain after 5, 10, and 15 days. [109]Cd did not accumulate in the middle of the endosperm in any stage of the grains, and the accumulation amount of [109]Cs decreased during seed development.

The crystallization of the rice grain develops during the maturing process. Therefore, the relation between the [109]Cd distribution and crystallization was investigated. It was found that along with the formation of crystallization in the grain (Fig. 6.4), the accumulation of [109]Cd in the endosperm decreased, and [109]Cd was confined at the surface of the grain. Although the image shows the distribution profile of [109]Cd and does not reflect the route of the movement of the tracer, the imbalance in the [109]Cd decrease on one side of the grain suggested that the transferring function of the grain was lost from this site.

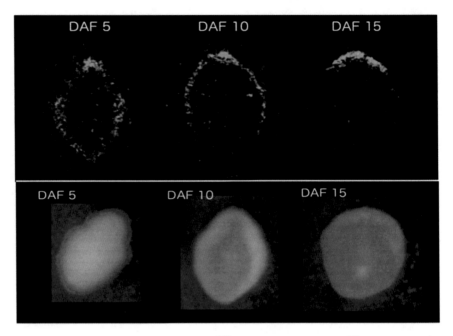

Fig. 6.4 Crystallization of a rice grain and the [109]Cd distribution [1]. Upper: [109]Cd distribution in a rice grain acquired by an IP; Lower: microscope image of a rice grain. During the ripening process of the grain, crystallization proceeds, and the color tone of the grain changes from an untransparent milky white to a partially transparent white color with the accumulation of [109]Cd to the surface and the decrease in the endosperm

6.2   3D Image of ^{137}Cs in a Rice Grain

In the case of ^{137}Cs, the rice plant was grown in water culture solution including ^{137}Cs (200 Bq/mL), and the grains were harvested 3, 5, 7, 9, 12, and 15 days after flowering. Then, each grain was sliced to 5 μm in thickness, and every 100 μm, successive slices were selected, in the same way as in the case of ^{109}Cd. Figure 6.5 shows the 3D image of the rice grain harvested 15 days after flowering. As shown in the figure, ^{137}Cs accumulated in the embryo and the outer bran layer. Before serving the grain as food, the periphery of endosperm is removed as bran through threshing. Therefore, the processed grain was estimated to contain a minimal amount of ^{137}Cs.

To compare the accumulation profile of ^{137}Cs with those of other elements in the grain, the distribution of K and Mg in the grain was analyzed by SEM-EDX (scanning electron microscopy/energy dispersive X-ray spectroscopy). Figure 6.6 shows the distribution of ^{137}Cs as well as of K and Mg in the horizontal plane of the grain. ^{137}Cs accumulated from an early stage of the maturing process of the grain, and the distribution was uniform throughout the grain. Distinct accumulation sites, including embryos, were not observed until the middle of the maturing process. Then, the accumulation of Cs at the grain surface and embryo began to increase,

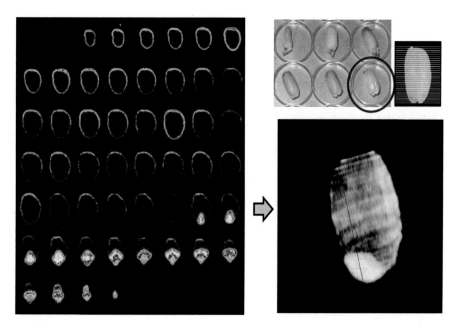

Fig. 6.5 Three-dimensional images of ^{137}Cs in brown rice. After 15 days of flowering, the rice grain was harvested, and the brown rice was sliced and exposed to an IP. Then, all images of the sliced sections were used to construct the 3D distribution of ^{137}Cs. The distribution of ^{137}Cs accumulated in the embryo, periphery of the endosperm, and surface of the brown rice is shown (most of these parts are removed as bran through thrashing before serving as a food staff)

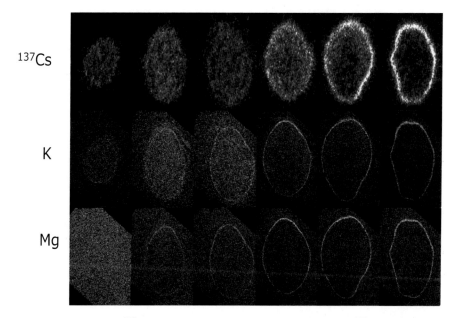

Fig. 6.6 Distribution of ^{137}Cs, K, and Mg in a rice grain. The distribution of ^{137}Cs, K, and Mg in the horizontal plane is shown during the development of a rice grain. The rice images were obtained 3, 5, 7, 9, 12, and 15 days after flowering from left to right. ^{137}Cs image was acquired by an IP. The K and M distributions were measured by scanning electron microscopy/energy dispersive X-ray spectroscopy (SEM-EDX)

along with a decrease in the Cs amount in the middle of the endosperm. On day 15 after flowering, the accumulation of Cs at the surface and embryo was completed. In the case of K, the distribution was uniform, similar to that of Cs, until the middle of the maturing process, and then the accumulation of K in the grain at the surface and embryo was observed. Since K and Cs are alkaline elements located in the same group in the periodic table, their chemical behavior was estimated to be similar. Magnesium was transferred to the grain later than K and Cs during the maturing process; however, the accumulation of Mg on the surface started from a rather early stage of maturation. The distribution of ^{137}Cs in the grain will be described in more detail in the next section.

Bibliography

1. Hirose A (2013) Ph.D. thesis. In: The University of Tokyo

Chapter 7
Microautoradiography (MAR)

Keywords Microautoradiography · Microautoradiography method improvement · ^{109}Cd · ^{33}P · ^{137}Cd · Microscopic distribution · Rice · Grain

7.1 MAR Method Developed

Microautoradiography (MAR) is a conventional imaging method based on the daguerreotype that was used to take photographs several decades ago. This technique was applied to visualize the distribution of radionuclide-labeled compounds within a tissue section. Although the distribution of radionuclides could be visualized in greater detail by MAR than by the IP (imaging plate) method because of the high resolution of MAR, IP imaging has become a widely used method because it has greater quantitative ability and sensitivity than MAR. For these reasons, MAR was performed only to obtain finer images and higher resolution of the radionuclide distribution. However, the application of the classical MAR method to plant tissue sections is associated with several difficulties, such as the preparation of thin sections. To overcome these difficulties, we developed a MAR method for imaging plant tissue section cells [1, 2, 3]. This method was applicable to fresh-frozen plant tissue and has two distinct features: (1) the sample was kept frozen from collected tissue for radioisotope detection, making it possible to fix solutes without solvent exchange, and (2) a 1.2 μm thick polyphenylene sulfide film was inserted between the fresh-frozen plant section and the photosensitive nuclear emulsion to keep the section separated from the emulsion before autoradiography was conducted, which significantly improved the quality of the section until microscopic detection, the quality of the MAR image, and the success rate.

There are two types of photosensitive nuclear emulsion preparations in MAR. In one, the photosensitive nuclear emulsion coats the plant section set on the slide to create a slide–section–emulsion layer. However, in this case, coated plant surfaces frequently exhibit irregularity due to the structure of plant tissue and produce a mixture of silver grains and plant tissue in the same microscope field. In our initial procedure, we attempted to improve this type of MAR in terms of overcoming the irregularity by optimizing the microscope manipulation, e.g. the composition of the

T. M. Nakanishi, *Novel Plant Imaging and Analysis*,
https://doi.org/10.1007/978-981-33-4992-6_7

microscope images sequentially obtained moving in the z-axial direction. However, the shadow of the plant cells and silver grains were not clearly distinguished, and the quality of the MAR images obtained thus far was insufficient for practical use.

In other preparation, the sliced section was mounted on photosensitive nuclear emulsion pasted on a glass surface, presenting a slide–emulsion–section layer and rarely resulting in irregularity. However, this type of MAR method had other problems, principally related to retention of the section during the development process of the emulsion film. In addition, the adhesive film for retaining the fresh-frozen section was found to hamper the autoradiographic procedure. In fact, we also attempted to develop an autoradiograph after the adhesive film was stripped from the sliced section sample while the sliced section remained on the emulsion surface. After several trials, the success rate of sample retention during the development process was found to be insufficient for routine usage.

The new method we developed achieved clear separation of the photosensitive nuclear emulsion and the section after exposure by sandwiching a 1.2 μm thick film between them (Fig. 7.1).

The emulsion on the glass slide and the section on the adhesive film were individually processed to develop the autoradiograph and tissue image, respectively. This procedure achieved high-quality image acquisition of the tissue section because the tissue section was rescued before the glass slide was exposed to the reagent for the development process. Other advantages of this method are an extremely high rate of sample retention and a wide choice of tissue staining. On the other hand, the disadvantage of this method is that the positioning accuracy could be reduced due to the separation of the section and the photosensitive nuclear emulsion. The insertion of the 1.2 μm thick film created a concern about the masking effect, particularly in detecting very low-energy β-rays. To address this point, a trial experiment was performed using ^{3}H, which emits β particles with a mean energy of only 5.7 keV, and obtained evidence that the β-ray could produce silver grains through the film (data not shown).

7.2 MAR of ^{109}Cd and ^{33}P in a Rice Plant

As an example of MAR images, the ^{109}Cd distribution of nonelongated stems and shoots above the stem of a rice plant was selected. The distribution images of ^{109}Cd and ^{33}P in the cross section of the crown root are shown.

Figure 7.2 shows the distribution of ^{109}Cd in the rice shoot after 24 h of ^{109}Cd absorption. When the light image and MAR image were superimposed, the xylem cells (Xy) were found to contain less ^{109}Cd. ^{109}Cd was more concentrated in the nodal vascular anastomoses (NVA) and the regular vascular bundles (Rv) than in the enlarged vascular bundles (Ev). The ^{109}Cd concentration in phloem cells was found to be particularly high, whereas the xylem cells contained only small amounts of ^{109}Cd. In the leaf sheath, the large vascular bundles produced more silver grains than the small vascular bundles. The ^{109}Cd MAR image could first provide evidence of

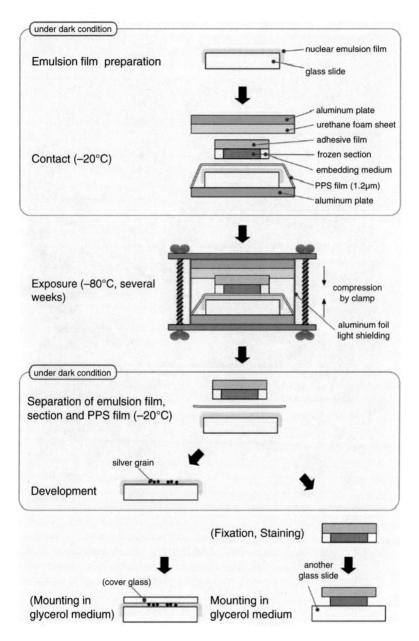

Fig. 7.1 Schematic illustration of the revised microautoradiography (MAR) method for fresh-frozen plant sections characterized by the separation process [1]. A 1.2-μm-thick polyphenylene sulfide (PPS) film was inserted between the frozen section sample and the photosensitive nuclear emulsion on the glass slide to easily separate the section and emulsion after several weeks of exposure. Progression of the development and tissue processing separately improved the likelihood of success of MAR, and fresh-frozen plant sections of 5 μm in thickness attached to the adhesive film were prepared

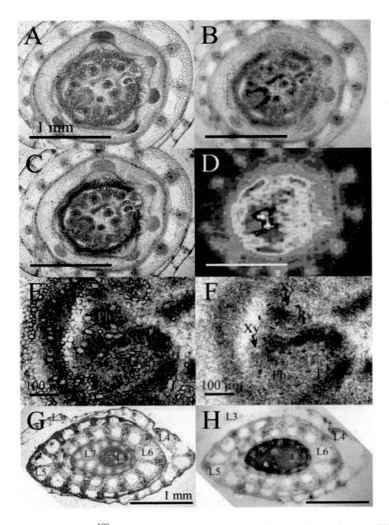

Fig. 7.2 Distribution of [109]Cd in rice, nonelongated stem and shoot above the stem [1]. (**a**) Transverse section of the nonelongated stem including the node stained with hematoxylin. (**b**) Autoradiograph of (**a**) showing the localization of [109]Cd. (**c**) Superimposed image of (**a**) and (**b**) after the RGB color in (**a**) was converted to grayscale and the *black color* in (**b**) was converted to *red*. (**d**) IP image of (**a**) with pseudocolor. (**e, f**) Magnified images of the node tissue corresponding to the open boxes in (**c**) and (**b**). (**g**) Transverse section of the part of the shoot above the stem stained with hematoxylin. The leaves were numbered according to their growth from L3 (old) to L8 (young). The first and second leaves (L1 and L2) were in senescence and depressed. (H) Autoradiograph of (G) of the [109]Cd accumulation in the young growing sink leaves, e.g. the seventh (L7) and eighth leaves (L8). Xy: xylem cells; NVA: nodal vascular anastomoses; Rv: regular vascular bundles; Ev: enlarged vascular bundles; Ph: phloem. The radioactivity of [109]Cd in the phloem (Ph) in the Rv was significantly higher than that in the Ev

the intensive accumulation of ^{109}Cd in the phloem in the nonelongated stem as early as 24 h after ^{109}Cd administration to the root. The xylem to phloem transport system mediating ^{109}Cd transport to the phloem in the nonelongated stem implied a significant impact on cadmium partitioning in rice plants.

With regard to the ^{109}Cd signals detected in these tissues using an IP, the signal intensity had a linear relationship with radioactivity. Although the MAR images were related to the ^{109}Cd concentrations, there was no linear relationship between the whiteness of the image and the ^{109}Cd amount, since there were sigmoidal dose–response curves, which are the principle of daguerreotype images. Another problem was that it was difficult to compare the blackening quantitatively across slides, since the precise thickness of the emulsion layer formed on the glass slide could differ among slides. When the stripping film with a uniform thickness was commercially available, the quantitative capability of MAR was ensured; however, the film is no longer in production. Therefore, the precise quantitative capability of MAR is now difficult to ensure.

Figures 7.3 and 7.4 show the distribution of ^{109}Cd and ^{33}P in the crown root of rice plants after 24 h and 15 min of absorption, with two and four metaxylem cells in the middle, respectively. As shown in Fig. 7.3, the ^{109}Cd signal intensity was low in the cortex tissue inside the sclerenchymatous (S) cell layer. A high silver grain image was observed at the outer circle overlapping with the epidermis, exodermis, and sclerenchyma cells, whereas the cortex tissue contained a small amount of ^{109}Cd, which was also observed in the IP image. Within the xylem cells in the stele, which had a high concentration of ^{109}Cd, the degree of blackening was similar to that in the surrounding cells, which could not be distinguished by the IP image. MAR clearly revealed the distribution of ^{109}Cd at the tissue level with high resolution.

Figure 7.4 shows the distribution of ^{33}P in the crown roots of rice plants after 15 min of absorption. Interestingly, ^{33}P-phosphate produced many silver grains on the sclerenchyma cells and their neighboring cells in the cortex, adjacent to the epidermis and exodermis cells, indicating that phosphate could easily pass through the epidermis and exodermis cell layers. Phosphate remained in the area of the sclerenchyma cells, which were expected to be connected with each other by Casparian strips, and then easily reached the stele without accumulating in the cortex. The success in visualizing ^{33}P after 15 min of exposure demonstrated both the characteristic transport mechanism of phosphate within the root tissue and the high sensitivity of autoradiography, suggesting the high potential applicability of MAR for ion uptake and transport analysis.

7.3   MAR of ^{137}Cs in a Rice Grain

The 3D ^{137}Cs image in a rice grain was shown in the previous chapter (Figs. 6.3 and 6.5), where ^{137}Cs accumulated in the embryo and the outer bran layer. To acquire a finer image of ^{137}Cs distribution in the grain, especially in the embryo, MAR was performed on rice grain containing ^{137}Cs. A rice plant was supplied with ^{137}Cs

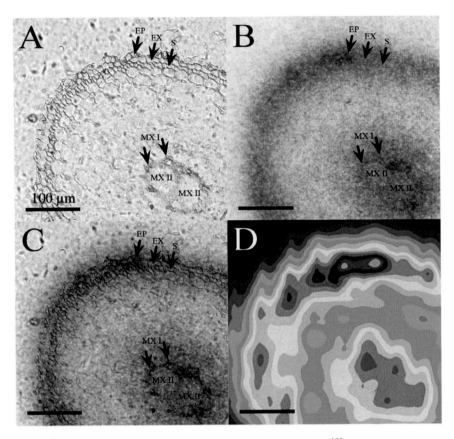

Fig. 7.3 Distribution of ^{109}Cd in the rice crown root after 24 h of ^{109}Cd absorption [1]. (**a**) Transverse section of the root with two metaxylem II (MX II) cells in the middle. (**b**) MAR of (**a**). (**c**) Autoradiograph superposed to (**a**). Red color was added. (**d**) IP image of the section adjacent to (**a**). Pseudocolor was added. The ^{109}Cd signal intensity was low in the cortex tissue inside the S cell layer. Arrows denote the root epidermis (EP), exodermis (EX), sclerenchymatous (S) cell layers, and metaxylem I and II (MX I and MX II) cells. The arrows denote the same site as Fig. 7.4. The ^{109}Cd signal intensity was low in the cortex tissue inside the S cell layer

solution (200 Bq/mL) for 24 h when flowering was started. Then, after 28 days, the mature brown rice was harvested, and MAR images were acquired. Figure 7.5 shows the distribution of ^{137}Cs in the embryo. As shown in the figure, most ^{137}Cs accumulated in the scutellum, and a very small amount of ^{137}Cs was observed in the plumule, radicle, and embryonic root, suggesting that the tissue with meristems that grew after germination was highly protected from the heavy element Cs.

To compare the distribution of Cs with those of the other elements in the embryo of brown rice, SEM/EDS (scanning electron microscopy/energy dispersive X-ray spectroscopy) was performed to acquire the distributions of K, Mg, N, P, Si, and Ca. As shown in the figure, K, Mg, and P showed higher concentrations at the

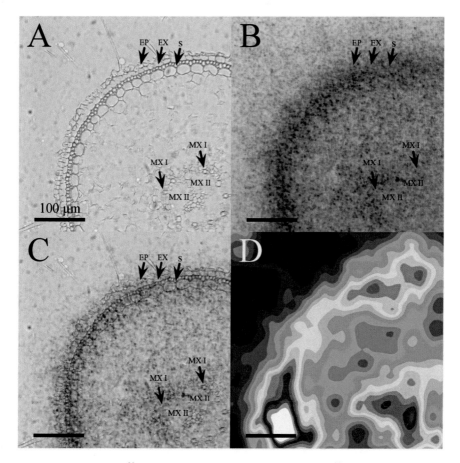

Fig. 7.4 Distribution of ^{33}P in the rice crown root after 15 minutes of ^{33}P-phosphate absorption [1]. (**a**) Transverse section of the root with four metaxylem II (MX II) cells in the middle. (**b**) Autoradiograph superposed to (**a**). Red color was added. (**c**) IP image of (**a**). Pseudocolor was added. The *arrows* denote the same site as Fig. 7.3. ^{33}P-phosphate concentrated in the sclerenchyma cells and their neighboring cells in the cortex, which were adjacent to the epidermis and exodermis cells

embryo; however, similar to ^{137}Cs, these elements were also concentrated at the scutellum, not at the plumule or radicle.

Since ^{137}Cs behavior in rice plants has attracted attention, especially after the Fukushima nuclear accident, the uptake behavior of ^{137}Cs was studied in comparison with that of K. The differences in the transport characteristics in plants between potassium (K^+) and cesium (Cs^+) were investigated using their radionuclides [42]K^+ and [137]Cs^+. The result is briefly introduced below. A tracer experiment using nutrient solutions supplied with [42]K and ^{137}Cs revealed that the ratio of the root's K^+ uptake rate to its Cs^+ uptake rate was 7–11 times higher than the K^+ to Cs^+ concentration ratio in the solution, and the number varied depending on the K concentration in the

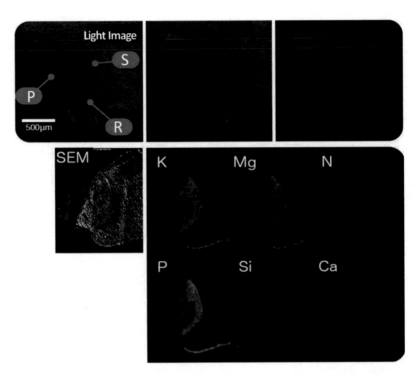

Fig. 7.5 Distribution of ^{137}Cs in the embryo of a rice train partially. *Red color* was added to the microautoradiography image (MAR) according to the intensity of the image. In the light image under the microscope, P: plumule, S: scutellum, R: radicle (embryonic root). The distributions of K, Mg, N, P, Si, and Ca are shown in grayscale. K, Mg, and P concentrations were high in the embryo; however, they were not accumulated in the plumule or radicle, which is similar to those of ^{137}Cs

solution as well as on the growth condition. After entering through the root tissues, the ^{42}K$^+$ to ^{137}Cs$^+$ ratio in the shoots was 4.3 times higher than the value in the roots. However, the ^{42}K$^+$ to ^{137}Cs$^+$ ratio in each leaf did not differ significantly, indicating that the primary transport of K$^+$ and Cs$^+$ in the shoots is similarly regulated. In contrast, among the radionuclides stored in the roots over 4 h, 30% of the ^{42}K$^+$ was transferred from the roots over the following hour, whereas only 8% of ^{137}Cs$^+$ was moved. In addition, within the xylem, K$^+$ was shown to travel slowly, whereas Cs$^+$ passed quickly through the roots into the shoots. Our study demonstrated the very different transport patterns for the two ions in the root tissue, employing ^{42}K$^+$ and ^{137}Cs$^+$.

Bibliography

1. Hirose A, Kobayashi NI, Tanoi K, Nakanishi TM (2014) A Microautoradiographic Method for Fresh-Frozen Sections to Reveal the Distribution of Radionuclides at the Cellular Level in Plants. Plant and Cell Physiology 55:1194–1202
2. Sugita R, Hirose A, Kobayashi NI, Tanoi K, Nakanishi TM (2016) Imaging techniques for radiocesium in soil and plants. In: Agricultural Implications of Fukushima Nuclear Accident. Nakanishi, T.M., Tanoi, K, pp 247–263
3. Hirose A (2013) Ph.D. thesis. In: The University of Tokyo

Chapter 8
Other Real-Time Movement

Keywords HARP camera · Circumnutation · Root movement · Moving angle · Movement cycle · AI

8.1 Root Movement During Growth (HARP Camera Images in the Dark)

Circumnutation is a rotating movement of a growing plant organ that is considered to result from endogenous rhythmic processes. The regular autonomous rotating movement of a root is due to different growth on the sides of the root, based on oscillators at the cellular or tissue level. However, the mechanisms responsible for circumnutation are still unclear. In particular, because of the lack of imaging tools, it has not been possible to acquire rotating images of roots under dark conditions. Since light has an influence on the physiological activity of the root, such as inducing photosynthesis, the development of devices that enable seeing the root under dark conditions is needed. Since we could use a super-HARP camera developed by the NHK broadcasting technology institute in Japan, we could visualize root circumnutation in the dark.

Four-day-old seedlings of rice plants were grown in water culture, and the super-HARP camera was set outside the phytotron to obtain images of the plant root through the glass window (Fig. 8.1). Images of the root were acquired at an interval of 1 min, and each image was accumulated for 4 s. The culturing room was kept dark, and the root-growing behavior was monitored from the next room. The rice image taken by the camera was analyzed by a computer to calculate the root length as well as root tip angle. Figure 8.2 shows an example image taken by the Super-HARP camera. Although the room was dark, the color image of the sample was shown with high resolution.

It was found that the curvature was initiated in the same region of the root, the elongation zone. The root elongation rate was kept constant with regular rhythmic

Supplementary Information The online version of this chapter (https://doi.org/10.1007/978-981-33-4992-6_8) contains supplementary material, which is available to authorized users.

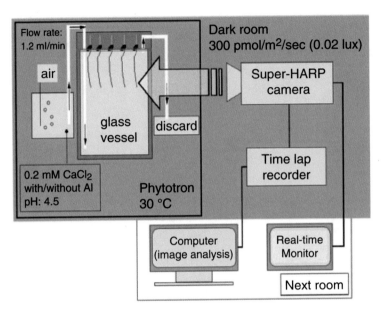

Fig. 8.1 Schematic illustration of the root imaging system using a Super-HARP camera in the dark [1]. A rice seedling was placed in the upper part of a glass vessel filled with 0.2 mM $CaCl_2$ solution. The root-growing manner was video-recorded by a Super-HARP camera, and all devices were maintained in the dark at approximately 300 pmol/m^2/s (0.02 lux). Since the room was kept in the dark, the root-growing manner was monitored from the next room

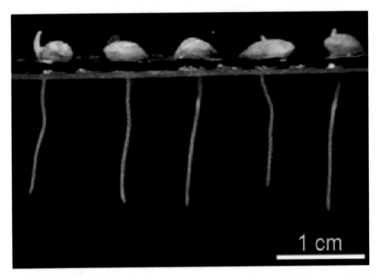

Fig. 8.2 An example of the rice root growing image taken by a Super-HARP camera [1]. The color image was taken at 1-min intervals and accumulated for 4 s

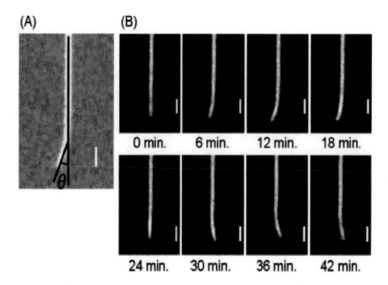

Fig. 8.3 Images of the circumnutation of a rice root by a Super-HARP camera [1]. (**a**) Measurement of the angle between the root tip and the direction of root elongation. (**b**) Successive root images grown in the control solution at pH 4.5, taken by a Super-HARP camera. The curvature was initiated in the same region of the root, which was the upper part of the elongation zone. The root length and rotation angle of the root tip were analyzed from the series of obtained images. The *white bar* is 1 mm in all images

movement, showing a fixed angle. The successive images of the root while growing in control solution are shown in Fig. 8.3. One round of root tip rotation took approximately 50 minutes.

However, when treated with Al solution, the rotation angle of the root tip was decreased, and then the movement resumed (Fig. 8.4). With increasing Al concentration, the time root rotation was suppressed, i.e., the time until the root resumed the rotating movement was increased. When the $AlCl_3$ concentration was increased to 50 μM, resumption of the rotating movement was not observed even after 7 h. With 5.0 μM Al treatment, inhibition of root tip circumnutation was obtained without inhibiting root elongation (Fig. 8.5). The results offered an interesting suggestion that this new inhibition of root tip circumnutation occurred at lower Al concentrations than the inhibition of root elongation previously reported in studies of Al toxicity. This is a new phase of early reactions of the plant root to Al stress.

It is known that Al ions induce callose production, as an earliest reaction, damaging plasma membranes and inhibiting root elongation. Our study suggested that the inhibition of root tip circumnutation occurred before callose production, followed by the inhibition of root elongation.

Whole root movement is controlled by the balance of plant hormones, and auxin probably plays a central role in root development because of its cell extension effect. Since auxin extends root cells, it has been supposed that the localization of auxin

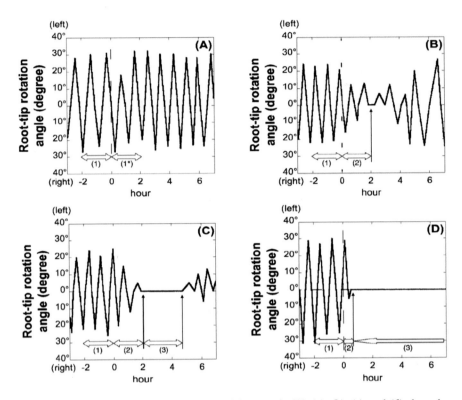

Fig. 8.4 Effect of Al on the circumnutation of the root tip [1]. (**a**), (**b**), (**c**), and (**d**) show the circumnutation profile when the solution was changed to 0, 5, 10, and 50 μM Al, respectively. (*1*) indicates the control period (2 h); (*1**) indicates the control period after renewal of the solution (2 h); (*2*) indicates the cycle-fading period; (*3*) indicates the cycle-ceased period. 0 h in each figure indicates when the solution was renewed. The rotation angle was calculated from the apparent maximum oscillation in the successive video images. With 5 μM AlCl$_3$, the maximum rotation angle decreased, and the root rotated again after a while without changing the elongation rate

generates circumnutation. As an alternative, some environmental conditions have been discussed, including the possibility that certain metal ions induce circumnutation.

Circumnutation plays an important role in the effective development of roots in soil structures containing inhomogeneous nutrients. With a rotating root tip, roots are able to select the appropriate direction to extend smoothly. Because of the lack of tools, it was extremely difficult to study root movement in the dark. The Super-HARP camera used in this study allowed the acquisition of images in dark conditions, approximately 300 pmol/m^2/s, under which no other optical devices could obtain images. Thus, the observation of root growth using the Super-HARP camera is better than any other method to study root activity in the dark.

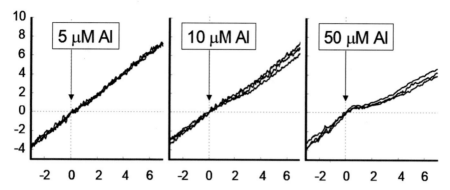

Fig. 8.5 Root elongation calculated through image analysis [1]. The root elongation was indicated by the relative root length, where 0 mm was set when an Al treatment was performed (at 0 h). With the 5.0-μM Al treatment, the root tip circumnutation was inhibited without inhibiting observable root growth, which suggests that the inhibition of circumnutation is an earlier effect than the inhibition of root elongation. The reason why circumnutation resumes after a time period is unknown

Bibliography

1. Hayashi Y, Nishiyama H, Tanoi K, Ohya T, Nihei N, Tanioka K, Nakanishi TM (2004) An aluminum influence on root circumnutation in dark revealed by a new super-HARP (high-gain avalanche rushing amorphous photoconductor) camera. Plant and Cell Physiology 45:351–356

Summary and Perspective

First, let us look again at the water imaging and movement presented. It was shown how an exquisite water-specific image could be shown by neutron beam imaging. It provided a beautiful profile of water in living plants. Since plant activity is rather slow, successive images of the plant acquired by the neutron beam could show slow movement of water that was not able to be visualized before. In the case of fast movement of water, outside our devices, it was noted that water circulation occurred in the internode. When the role of water is discussed, focus is often on water absorption or the movement of newly absorbed water, rather than the water already present in the plant. One of the points suggested by our research was that both newly absorbed water and the water already present in the plant seemed to have their own role. When water was labeled with ^{15}O, whose half-life is extremely short, 2 min, the trace of water movement could be measured by our measuring system. What the ^{15}O-water movement showed was that a tremendous amount of newly absorbed water was constantly flowing out from the xylem tissue horizontally and pushing the water already present into the xylem, where it was transferred upwards. However, the water flow in the xylem tissue was kept constant. Is there any difference between the newly absorbed water and the water already present in the tissue? How could the newly absorbed water be distinguished from the water already present in the internode? It was the first measurement of water circulation within the internode.

Although we could measure the water circulation movement only in the internode, there might be other circulation movements in other tissues. If this is the case, then there is a further question of why there is a circulation movement of water within the internode. Since the water movement was different from that of the dissolved ions in it, there must be an important role for the water movement itself. This movement might be changed during the developmental stage of the plant, which could be estimated from the development of bellows-like lignin around the xylem tissue. The pitch of the lignin bellows is large during the young stage and allows water to flow out of the xylem easily; however, the pitch gradually becomes narrower towards the senescent phase and hardly allows water to leak out at the end. It was suggested that the amount of water leaking out from the xylem was adjusted

© The Author(s) 2021
T. M. Nakanishi, *Novel Plant Imaging and Analysis*,
https://doi.org/10.1007/978-981-33-4992-6

by this lignin, that is, the young tissue contains a high amount of water, and the amount of water reaches a minimum at the senescent phase. It was also estimated that the development from the juvenile to the adult phase might be created from the degree of water circulation activity, not only the amount of water but also its movement and speed.

The water image of the rape plant pod was acquired to study whether the water flow changed the phase of the developmental stage of the seed. When the water amount was measured by harvesting the seeds in the pod, it was found to increase before oil formation started, and then, the water content decreased with increasing oil formation. Although the change in the water amount in the seed stem could not be distinguished because of the fine structure, it was suggested that the change in the water flow might trigger oil formation in seeds.

When water is present, biochemical reactions proceed, and the plant can grow to the next stage. However, if it is possible to regulate the water flow, especially when it is possible to stop a specific water flow, this could be a survival strategy for plants. Considering the environmental conditions of the plant, biochemical reactions could cease if the water supply was stopped. As a result, without further futile use of energy, the plant can sterilize a specific tissue for further development. The water movement in a rose flower causing the bent neck phenomenon shown by neutron beam imaging suggested this possibility for the water regulation strategy by the plant.

Whether there is enough water in the environment is an important factor for plant survival; however, only water absorption activity, high or low, tends to attract attention to study the drought tolerance of plants. As shown in our study, naturally developed drought-tolerant cowpea usually absorbed less water than sensitive cowpea. However, under drought conditions, tolerant plants began to absorb much higher amounts of water, whereas sensitive plants could not absorb water. This result was unexpected. Generally, we expect drought-tolerant plants to have higher water absorption activity than sensitive plants, which is why tolerant plants can survive under semi-arid conditions. Therefore, it was natural for us to study the mechanism of water absorption and try to introduce strong water absorption activity to the sensitive plant. However, our results in naturally created drought-tolerant and drought-sensitive plants suggested that the drought tolerance activity might be related to water movement within the plant. It might be that the drought-tolerant plants utilized a small amount of water more effectively than the sensitive plants under normal conditions, which might be derived from the different water movement activities. However, it is not known how tolerant plants can absorb more water than before under drought conditions.

Regarding the velocity of water absorption, there are many questions that cannot be solved. One of them is what the normal speed of water absorption is. As shown in the comparison of water culture and soil culture, the plant grows much faster in water culture. In the case of plant factories, water culture is generally employed to grow vegetables, since plants grow much faster in water culture than in soil culture. However, in cereal plants, the amount of grain produced, the yield, is much lower in water culture than in soil culture. Therefore, we have to depend on soils to grow cereal plants, such as wheat, corn, or rice, even though the growth is slow. Considering the drastic difference in yield between water and soil culture, one of our expectations was that it might also be related to water circulation in the plant. In water culture, the plant grows too fast, especially during the juvenile phase, and might be unable to regulate water movement.

The root tip moves during growth, known as circumnutation. Because of this movement, the soil is pushed aside to facilitate root growth and guide the orientation of the growth direction. As a result, this movement makes a space adjacent to the root surface in soil. As is known, soil contains roughly similar amounts of soil matrix, water, and air. Therefore, combined with circumnutation, it is conceivable that there could be an air space close to the surface of the root.

The neutron imaging clearly showed how the roots were growing in soil and the water amount in the vicinity of the root and revealed that there was hardly any water solution touching with the root surface. The neutron image of the root imbedded in soil suggested that the root was absorbing water vapor, not water solution. Then, what about the metals is the next question. Is the root absorbing metal vapor? We have no knowledge of what chemical form of water or elements the roots are absorbing.

In the case of root movement, a very interesting phenomenon was found using the Super-HARP camera, which enabled the visualization of root movement in the dark. When there was a chemical change in the environment, although the circumnutation of the root tip ceased, the root was able to elongate, and it was interesting that after a while the root movement resumed. Although the first data on plants show that the harmful effect is growth inhibition, the first effect of the toxicity was to stop the rotation movement of the roots before growth inhibition occurred. In the case of a rice root, one round of movement of the rice root tip showed a constant time of approximately 50 minutes. However, this movement ceased when Al ion was supplied. Al ion is known to be toxic for plant growth. When the Al concentration was low, the root could grow even if the circumnutation ceased and resumed after a while. However, the roots could not grow or resume movement when the concentration of Al was increased. The time needed for resuming the movement of the root tip was dependent on the Al ion concentration. It is not known what triggers the resumption of the movement of the root tip. This visualization of the root movement indicated that it is very important to consider plants as moving, not immobile organisms. Immobility and mobility are often judged only out of our sense of time scale. It seemed that there was a regulatory system of root movement that might be related to animal movement.

The concentration profile of each element in plant tissue was found to spread systematically throughout the developmental stage of the plant by neutron activation analysis (NAA), which allows nondestructive multielement analysis and is the only method to measure the absolute amount of the elements. The features of the concentration profile in a plant differed from element to element, and each element concentration showed differences between tissues. This concentration gap was also element specific and occurred throughout the plant. These profiles of the concentrations of the elements and their concentration gaps seemed to regulate the activity of each tissue. For example, there was a diurnal change in Mg and Ca concentrations, especially in the apical meristem. However, Mg-specific features have not been well studied because of the lack of tools to separate Mg behavior from Ca behavior. There is no suitable RI to trace Mg movement, and Mg cannot be visualized by staining because Mg is always superposed by an overwhelming amount of Ca. We tried to produce ^{28}Mg, whose half-life is 21 h, by the ^{27}Al(α, 3p)^{28}Mg reaction and applied ^{28}Mg as a tracer for the first time in a plant study. The properties of Mg were gradually revealed, and the study is currently under development, but the role of Mg in maintaining the homeostasis of the plant has not yet been successfully clarified.

Another element with an interesting diurnal change in concentration is Al. Since the sensitivity of NAA to Al is extremely high, the trace amount of Al contained in a morning glory seedling was measured. A ng level of Al was regularly secreted from the root tip, and the amount of Al outflow decreased with time. The movement of this Al was not due to artificially added Al but to Al already contained in the seed. This secretion movement at the root tip was another interesting phenomenon suggesting that there might be a rhythm in root tip activity.

From the features of the elemental profile, not only the plant itself but also the elemental conditions of the environment can be analyzed. Plants acquired methods to adapt to environmental conditions through their long history of evolution. Some of them evolved to survive under high concentrations of toxic elements, such as Se. On the other hand, when the same plants were grown in different districts, the elemental profiles could be different, reflecting the features of the soil. The agricultural production district could be identified by analyzing the trace elements absorbed in the products. Not only the plants but also animals fed with the plants grown in different districts showed elemental profiles corresponding to the soil where the plants grew.

The real-time RI imaging system (RRIS) was developed to visualize and analyze how each element is absorbed and moves within the plant. As shown, each element had its own specific movement and accumulation pattern when absorbed by the plant. Comparing these patterns with the movement of water raised the question of why the movement of each element could be so different from that of water, showing its own movement pattern, and how the movement of the elements dissolved in water could be regulated differently. There must be many kinds of different transporters for the elements at many different sites in the plant. Even if we could identify these transporters, it still seems impossible to know why the elements are moving or what triggers each transporter to maintain different activity to maintain homeostasis of the plant. These results seem to suggest that to understand these features, the activity of the whole plant should be studied at the same time.

Another interesting result shown by the RRIS was that the element movement in the water or soil around the root could be visualized. In the case of water culture, when ^{14}C-glutamine was supplied to the culture solution, the ^{14}C signal accumulated at a certain distance from the rice root tips, and then, when the concentration of glutamine was high, all of the ^{14}C signal suddenly moved together to the roots to be absorbed. The absorption of ^{14}C-glutamine did not occur constantly but with a rhythm. Until this visualization, it had never been thought that there was a method to visualize real-time chemical reactions in water, which many chemists are interested to see.

In the case of soil culture, for example, when ^{32}P-phosphate was supplied, the site in the soil from which the root absorbed the phosphate could be visualized. Since phosphate is adsorbed from soil, the root could absorb only the phosphate close to the root, which resulted in a depleted image of ^{32}P-phosphate in soil as an enlarged shape corresponding to the roots.

The physiology of plants growing in soil has not been well studied. One of the reasons is that soil itself has a very complicated structure and function; therefore, it is difficult to discuss the plants growing in this complicated soil. The water absorbing activity of roots in soil is also not well understood. The difference in the movement of water from the movement patterns of nutrients suggested a large number of possible hypotheses. For example, only the amount of water absorption could be regulated to a low level while enough nutrients were supplied, it might cause a change from the juvenile phase to the adult phase, and it could be one of the solutions to growing cereal plants in a plant factory.

Since the speed of the movement was different among the elements, as a result, the accumulation profile of each element was shown to be different. Most of the heavy elements accumulated in roots, except for Cr and Mn, and were not transferred to the aboveground part of the plant. Why heavy elements are absorbed is the next question. When they are absorbed and stay in the roots, there must be some role of the heavy elements only in the roots.

In the case of the main essential elements, when transferred to the aboveground part, they were first transferred to the youngest tissue. After enough of an element accumulates in the meristems, it moves to the other tissue by phloem flow.

When the microautography (MAR) method was modified and the distribution of heavy elements in the grain was visualized, they did not accumulate in the tissue to be grown as meristems, suggesting protection of the next generation from contamination.

The visualization of the carbon fixation process using $^{14}CO_2$ gas was very interesting. First, invisible CO_2 gas could be visualized as $^{14}CO_2$ gas in the air, and then, the gas was fixed by the plant tissue. This fixed carbon metabolite, photosynthate, was visualized and observed to transfer to the meristems and create new tissue, which means that the whole process of photosynthesis and the proliferation to produce the new tissue could be visualized. Since the main carbon source in plants is from CO_2 gas in the air, the visualization of the carbon assimilation process meant that we could trace the whole creation process of the plant structure.

Moreover, it was noted that the orientation of the photosynthate movement was different according to the tissue where CO_2 gas was fixed. In the case of Arabidopsis, as presented, the carbon fixed at the rosette leaves was transferred to the root and the main stem, whereas the carbon fixed in tissues other than the rosette leaves was transferred to the branch internode and hardly moved to the root. However, the route changed according to the developmental stage. As the developmental stage progressed, the rosette leaves supported the young stem, not the old main stem. It is not known how the route of photosynthate movement is regulated or how the orientation of phloem flow is controlled. The image analysis of the $^{14}CO_2$ gas fixation process enabled us to analyze the phloem flow in detail. For example, even carbon fixed in the small leaves in the branch stem, close to the meristems, was transferred to other tissues after enough carbon was transferred to the meristem. In the case of pod formation in a soybean plant, most of the photosynthate was supplied by the closest trifoliate leaves. Photosynthate movement is crucial to the development of new tissue, and photosynthate movement must reveal the most effective way to create the tissue.

Radioisotope imaging provides specific and direct imaging possibilities for many ions where no alternative solution with fluorescent probes exists. The RRIS offers quantitative and nondestructive access to mineral nutrition, from the short term to the long term, including the developmental stage of the plants. As shown, the RRIS offers a broad range of applications, especially in combination with fluorescent imaging.

Last, I want to reiterate that radiation and radioisotopes are indispensable and promising tools to trace water and element movement in plants and provide many fundamental but new questions to be studied. I also hope that these newly introduced methods of in vivo nondestructive imaging or measurement might open a new field of plant research, not only to reveal new functions or to evaluate intact systems but also to find ways to bridge the microscopic world of living plants with the macroscopic world.

Printed in the United States
by Baker & Taylor Publisher Services